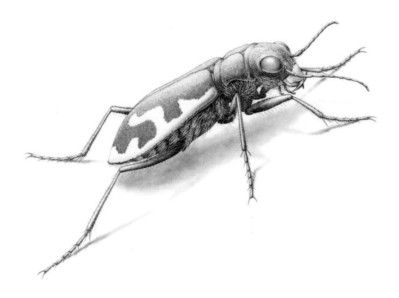

Illustrated by Peter Eades

The Pleasures of Entomology

Portraits of Insects and
the People Who Study Them

Howard Ensign Evans

Smithsonian Institution Press
Washington, D.C., 1985

©1985 by Smithsonian Institution
Second printing, 1986
All rights reserved
Printed in the United States of America

Designed by Carol Hare Beehler

Library of Congress Cataloging in Publication Data

Evans, Howard Ensign.
The pleasures of entomology.
Bibliography: p.
Includes index.
Supt. of Docs. no. : SI 1.2:En8/3
1. Insects. 2. Entomologists.
I. Title.
QL463.E932 1985   595.7   84-600318
ISBN 0-87474-421-0 (alk. paper)

The paper in this book meets the guidelines
for permanence and durability of the
Committee on Production Guidelines
for Book Longevity of the
Council on Library Resources.

Contents

To Mary Alice

Preface

If insects were the size of birds, or people the size of mice, "bug watchers" would be as prevalent as bird watchers, and entomologists would command the budget of the Defense Department. But as it is, entomologists have a good deal of trouble explaining what their science is all about, or for that matter how it is spelled. They are able to spout figures on the impact of insects on our economy and on human health, but in a world preoccupied with the high technologies of communication and of mass destruction, their voices are little heard. But this book is not an effort to extol the importance of insects and of entomologists; it is an attempt to persuade you that insects are fun to study and to contemplate. To those who are surfeited with the crazy drama of human events—or frightened by it—I especially recommend the insect. The insect never ceases to astound with its capacity to combine beauty and ugliness, power and frailty, familiarity and otherworldliness—all in a package so small we can eliminate it, if we will, with a stamp of a foot. But no fear, there will be others to take its place.

For critical comments and helpful suggestions, I am indebted to Kevin O'Neill, John Alcock, Barbara Mattingly, and Mary Alice Evans. And for inspiration, to all the insects I have known, far too numerous to mention by name. You'll meet a few of them in the pages that follow.

1
The Pleasures of Entomology

Entomologists are the most fortunate of people. They are students of the largest and most diverse group of living things on earth, the insects. To be able to understand and appreciate the cockroach, the bumble bee, the louse, and all the myriads of other six-legged creatures—what better way is there to enjoy the richness of the earth? Entomologists are also asked to control insects, and despite their admiration for them they do appreciate that insects unfettered sometimes become unduly competitive with the human species. Swatting a mosquito is allowable, considering that mosquitoes are so well able to flood the earth with their kind. But swat it respectfully; the mosquito is a product of millions of years of evolution, and a marvelous creature it is, equipped with its own hypodermic and as ready to sample the blood of a pauper as of a millionaire.

It is true that we rarely extend our sympathy for animals beyond the furry or feathery kind. The dim eyes of a caterpillar do not move us the way those of a rabbit do, and there will never be a Beatrix Potter of the insect world. Those who speak of animal rights do not concern themselves with insects, even though Shakespeare assures us that

The poor beetle that we tread upon,
In corporal sufferance finds a pang as great
As when a giant dies.

The entomologist, as everyone knows, is a rather wild-eyed individual who cares little for economics or politics. As William Kirby and William Spence wrote in 1815, "in the minds of most men . . . an entomologist . . . is synonymous with everything futile and childish." Why should anyone, they ask, become involved in a "science which promises only to signalize him as an object of pity or contempt in nine out of ten companies with which he may associate?"—a question they answered more amply, perhaps, than anyone has answered since. Kirby and Spence's *Introduction to Entomology* was the first major book in English on insects, preceded only by Thomas Mouffet's *Theatre of Insects* (1658), a translation from the Latin of a Renaissance insect bestiary. It was Mouffet's daughter, by the way, who is said to have inspired the immortal lines:

Little Miss Mouffet
Sat on a tuffet
Eating her curds and whey.
Along came a spider
And sat down beside her
And frightened Miss Mouffet away.

Kirby and Spence's book was dedicated to Sir Joseph Banks, "president of the Royal Society, etc.," and was an attempt to point out "the connection that exists between natural science, and agriculture, and the arts." It is a shame that so many books, the result of great skill and labor on the part of the authors, become no more than dusty tomes in back corridors of the library. Kirby and Spence are still good reading:

Insects appear to have been nature's favourite productions, in which . . . she has combined and concentrated almost all that is either beautiful and graceful, interesting and alluring, or curious and singular, in every other class and order of her children. . . . The sight indeed of a well-stored cabinet of insects will bring before every beholder not conversant with them, forms in endless variety, which before he would not have thought it possible could exist in nature, resembling nothing that the other departments of the animal kingdom exhibit, and exceeding even the wildest fictions of the most fertile imaginations.

"With respect to religious instruction, insects are far from unprofitable. . . ." William Kirby was for seventy years a village cleric, the rector of Barham, a village in Suffolk some seventy miles from London. One imagines him somewhat in the mold of the Reverend Septimus Harding, in Anthony Trollope's *Barchester Towers*. But with a difference—Kirby was intrigued by insects. It is said that he went about his duties with a bottle of gin in his hip pocket—not for himself, heaven forbid, but as a means of preserving the beetles he collected along the way. When Kirby was young, the fame of the Swedish naturalist Linnaeus was at its height, and when the Linnaean Society of London was founded in 1788, ten years after Linnaeus's death, Kirby was a charter member. Later in life he corresponded with the pioneer American entomologists William D. Peck, Thomas Say, and Thaddeus Harris. With Harris, Kirby felt a special bond, for Harris's father was a cleric and author of *The Natural History of the Bible*.

William Kirby was forty-six when he met another collector of beetles, William Spence. Together they planned their *Introduction to Entomology*, which came out in four volumes over the period 1815–1826. Eventually, it went through several editions and brought fame and modest fortune to its authors. More important, it elevated entomology to the status of a respectable science. A little over a century before, the naturalist John Ray had been asked to testify in court as to the sanity of Lady Granville, who collected butterflies. Now, it seemed, half the population of Britain was collecting insects!

Young Charles Darwin was then at Cambridge University, ostensibly training for the clergy. "But," he says in his *Autobiography*, "no pursuit at Cambridge was followed with nearly so much eagerness or gave me so much pleasure as collecting beetles. . . . I will give proof of my zeal: one day, on tearing off some bark, I saw a third and new kind, which I could not bear to lose, so that I popped the one which I held in my right hand into my mouth. Alas! it ejected some intensely acrid fluid, which burnt my tongue so that I was forced to spit the beetle out, which was lost, as was the third one."

Years later, the Staten Island naturalist William T. Davis told a similar story. Davis was a businessman and an amateur entomologist—in the sense that he pursued insects for love and not profit, like Kirby, Darwin, and so many other naturalists. Davis's biographer, Mabel Abbott, tells his story in these words:

> *On their way home [by train from Florida] their car got a hot box and the train had to halt. Davis and his friends got out to stretch their legs and, of course, looked around for insects. Davis turned over a board and found a cricket; there was also an elater [click beetle] which they had not seen before. . . . Davis took both insects—and at that instant the train began to move. He could not possibly bottle the bugs, and unfortunately he was also burdened with his camera. With both hands full, he ran frantically after the train, which was gaining speed. But he had to have one hand free to get on the car. There was not a second to spare. He popped the elater into his mouth, Lutz [his companion, Frank Lutz of the American Museum of Natural History] grabbed the other arm, and he landed on the step, with the elater safe as Jonah. It could have pinched him severely, but it didn't. Probably it was too surprised.*

Had Davis read Darwin's *Autobiography* and was his quick decision influenced by it? We shall never know. Incidentally, Alfred Russel Wallace, who proposed the concept of evolution by natural selection jointly with Darwin, at a famous meeting of the Linnaean Society of London on July 1, 1858, was also a noted collector of insects. Wallace was originally a teacher of English and drawing, but when a young hosier's apprentice named Henry Walter Bates showed him his cases of beetles and butterflies, his life was changed. Both were ready to cast aside their mundane lives, and together, in 1848, Wallace and Bates sailed for Brazil, planning to finance their trip by selling insects and other specimens to collectors.

Wallace eventually became the father of biogeography, and Bates the author of one of the classics of biology, *The Naturalist on the River Amazon*, as well as the originator of the concept of mimicry. Bates had discovered that certain butterflies that were distasteful to birds were closely copied in color, form, and behavior

---

Is there a better way to spend a summer's day?

P. EADES

by other, more palatable butterflies. Darwin found Bates's discussion of this now familiar concept a supreme example of the operation of natural selection, and included a section on mimicry in later editions of *The Origin of Species*.

Many a young person has found pleasure in collecting insects. A few have gone on to dedicate their lives to entomology; others have been led on to other life phenomena, and have become geneticists, physicians, or whatever; still others have gone into other fields entirely—but often with a secret longing for the joys of their youth. Winston Churchill, late in his life, planned a "butterfly garden," planted so that certain butterflies could breed successfully there year after year. Vladimir Nabokov's stories are sprinkled with butterflies and with entomologists, and Nabokov himself was an amateur lepidopterist. Some years ago, when I was entrusted with the care of the insect collections at Harvard's Museum of Comparative Zoology, I was several times approached by students who were writing Ph.D. theses on Nabokov. They wanted to see some of his butterflies, which had been donated to Harvard. On one occasion we had a visit from Nabokov himself. At that point I had read none of his novels and could not understand why our secretary was so awe-stricken (she had).

There is much to be said for collecting insects, either as a hobby or as a contribution to knowledge of insect variety and distribution. Even today, new kinds of insects are being discovered, particularly in the tropics—indeed, some estimate that not more than half the species of insects on earth have been found and described. But there is much more pleasure—and value—to be gained by studying the ways of living insects. Kirby and Spence were well aware of this.

> *You must leave the dead to visit the living [they urged]; you must behold insects when full of life and activity, engaged in their several employments, practicing their various arts, pursuing their amours, and preparing habitations for their progeny; you must notice the laying and kind of their eggs, their wonderful metamorphoses, their instincts, whether they be solitary or gregarious, and the other miracles of their history—all of which will open to you a richer mine of amusement and instruction, I speak it without hesitation, than any other department of Natural History can furnish.*

All of this is, of course, a good deal more challenging than the mere placing of a beetle in a bottle and later arranging it among other pretty specimens. When all the insect species have been catalogued (if this ever happens) we will still have much to learn about the roles they play in nature. It is safe to say that everyone, unless he lives in a thoroughly sanitized high-rise apartment building, has insects about him that are not fully understood. Those most homely of insects, the cockroaches, are the subject of intensive research in several laboratories. The yellow jackets that haunt September picnics are similarly keeping several researchers busy, and only a few years ago a yellow jacket species wholly new to science was discovered in the eastern United States. Even the honey bee, after centuries of close association with man, still has its secrets.

Fortunately, there are a few individualists who find the world of insects sufficiently absorbing that they don't much care if people sometimes regard them askance. They are the lucky ones, and they know it. They have the fun of watching incredible performances by creatures built very differently from us and meeting life's problems in quite different ways. They may not know the latest Dow Jones industrial average, but they know a good many things alien to Wall Street. And they believe that in the long run what they know may have more to do with survival in the real world.

When Kirby and Spence wrote their book, a good deal was already known about the ways of insects—otherwise they could scarcely have filled four volumes. From 1734 to 1742 six volumes had appeared in France, entitled *Mémoires pour Servir à l'Histoire Naturelle des Insectes*. The author was a self-sufficient amateur scientist, René Antoine Ferchault de Réaumur. Trained in law, mathematics, and physics, Réaumur made major contributions to subjects as diverse as metallurgy, meteorology, and the physiology of digestion. He was fifty-one when the first volume of his proposed ten-volume treatise on insects appeared. Although he lived to be seventy-four, he never completed the last four volumes. Volume seven, on ants, was unearthed in incomplete form in the archives of the French Academy of Sciences by Harvard professor William Morton Wheeler and was published in 1926.

Réaumur was in many ways the father of the science of insect behavior. As he put it:

> *The history of the insects is a great, I might say a vast domain, which may be traversed by many paths. The portion most interesting to me is that to which the majority of people are most susceptible, namely, the portion comprising everything relating to the engineering, the habits, in a word, to the industries of these small animals.*

He was by no means attracted to the art of classification (one suspects he never made an insect collection). At one point he wrote:

> *I confess that I am not in the least inclined towards a precise enumeration of every kind of insect, even if it could be undertaken. It seems to me sufficient to consider those kinds which prove to us that they deserve to be distinguished. . . . It seems to me that the many hundreds and hundreds of species of gnats and very small moths which exhibit nothing more remarkable than a few slight differences in the form of the wings or the legs, or varieties of coloration or of different patterns of the same colors, may be left confounded with one another.*

Réaumur would be shocked to learn that some of these "gnats and very small moths" have turned out to have very different behaviors! But for his time he had a remarkable sense of the need for discovering and describing the intricacies of insect behavior, both to satisfy one's curiosity and to provide knowledge of potential value to mankind. In this he contrasted sharply with the then dominant figure in French natural history, Buffon, who scoffed that "a bee should not occupy more space in the head of a naturalist than it does in nature."

Réaumur had indeed written about bees (and many other insects) at great length. He had observed honey bees in observation hives, between panes of glass, and in this way he was able to correct many false impressions about the functioning of the colony. He found that when the queen was removed, for example, the workers could produce a new queen by expanding a cell and feeding the larva on royal jelly. He studied food exchange and the functioning of the sting. He censused a hive by plunging it into cold water to immobilize the bees; then, with the help of his patient secretary, he counted 26,486 workers, 700 drones, and one queen.

Réaumur had several disciples, among whom was François Huber, who was blind from youth but nevertheless devoted his life to the study of honey bees, assisted by his servant and later by his son Pierre. The elder Huber described the mating flight and the mode of ventilating the hive, among other things. Pierre adopted an experimental approach that was well ahead of its time. His comment that "each species has its own habits, each individual its own constitution" also has a modern ring. It was Pierre Huber who discovered that some species of ants enslave others, a phenomenon confirmed in England by Charles Darwin, who discussed it at some length in *The Origin of Species*.

The French language may be especially effective in conveying the goings-on in the insect world; perhaps its mellifluousness does better justice to the hummings and scrapings of insects than do the harsher accents of English and German. It was an essay on the behavior of a hunting wasp by a retired army surgeon, Léon Dufour, that inspired a young French school teacher to devote his life to studying the lives of insects, one who has inspired many another. Jean Henri Fabre described his reaction to Dufour's essay in these words:

> *New lights burst forth: I received a sort of mental revelation. So there was more in science than the arranging of pretty beetles in a cork box and giving them names and classifying them; there was something much finer: a close and loving study of insect life.*

Fabre was the son of an illiterate mother and a father who was a chronic failure as an innkeeper. Nevertheless, he struggled to obtain an education and eventually became a professor at the Lycée at Avignon. Here he taught for nearly twenty years at a meager salary. When he was eventually discharged for his unconventional ideas on teaching (such as admitting girls to his science classes), he was forced to support his wife and several children by writing. One result was *Souvenirs Entomologiques*, which appeared in ten volumes, from 1879 to 1907. These were soon translated into English by Alexander Teixiera de Mattos under such titles as *The Hunting Wasp*, *The Life of the Grasshopper*, and *The Glow-worm and Other Beetles*. Like Réaumur, Fabre had insufficient regard for the identification of the insects he studied, and, working in obscurity as he did, he had little opportunity to discuss his findings with others. He had little use for the concept of evolution, although Darwin spoke of him as an "incomparable observer." But his enthusiasm, his sense of adventure, his love of insects sparkle from every page of his writings. In his later years he asked:

> *Is it really worth while to spend time . . . in gleaning facts of indifferent moment of highly contestable utility? Is it not childish to enquire so minutely into an insect's actions? Too many interests of a graver kind hold us in their grasp to leave leisure for these amusements. That is how the harsh experience of age impels us to speak: that is how I should conclude, as I bring my investigations to a close, if I did not perceive, amid the chaos of my observations, a few gleams of light touching on the loftiest problems we are privileged to discuss.*
>
> *What is life? . . . What is human intelligence? What is instinct? Are these two mental aptitudes irreducible, or can they both be traced back to a common factor? . . . The theorists, proudly daring, have an answer nowadays for every question; but as a thousand theoretical views are not worth a single fact, thinkers untrammeled by preconceived ideas are far from becoming convinced. Problems such as these, whether their scientific solution be possible or not, require an enormous mass of well-established data, to which entomology, despite its humble province, can contribute a quota of some value. And that is why I am an observer, why, above all, I am an experimenter.*
>
> *It is something to observe; but it is not enough: we must experiment, that is to say, we must ourselves intervene and create artificial conditions which oblige the animal to reveal to us what it would not tell if left to the normal course of events.*

Late in his long life, Fabre was discovered by the literary world, and praised by Romain Rolland, Edmond Rostand, and Maurice Maeterlinck. It was the popular Belgian poet and playwright Maeterlinck, more than any other person, who brought the fascination of insects to broad public attention in books such as *The Life of the Bee*, *The Life of the White Ant*, and *The Life of the Ant*. Maeterlinck not only read widely but studied these insects himself. He described their behavior in a style larded with philosophy and with anthropomorphisms. Termites, for example, were said to lead a "ferocious, incarcerated life, barbarous, furtive, and merciless." Like so many early naturalists, Maeterlinck is still good reading if taken in the context of his time and of the audience he was addressing. He sought the source of altruistic behavior in animal and human societies, a matter still being bandied about (one hopes with greater success) by the sociobiologists of today.

Fabre left a dual legacy. He wrote in a personal, anecdotal style for a general audience, setting the stage for the more erudite Maeterlinck and for many others, in our time especially for the late Edwin Way Teale. The experiments he performed in the field also presaged those of many a later worker, for example for Niko Tinbergen's elegant studies of the hunting and orientation of beewolves. The experimenter *par excellence* was, of course, Karl von Frisch, and when he, Tinbergen, and Konrad Lorenz shared the Nobel Prize in physiology in 1973, the study of insects—and of animal behavior generally—may be said to have passed a milestone.

There is in fact no evidence that von Frisch had been influenced by Fabre, although there are parallels—and contrasts—in their lives. Karl von Frisch came from a family of scientists and was trained at the universities in Vienna and Munich; he exchanged ideas with biologists throughout the world, and his experiments were far more exhaustive and sophisticated than those of Fabre. The problems he had to overcome were also very different, most notably the distress and destruction of two world wars, the second of which destroyed both his home and his laboratory. But like Fabre he felt the need "to write so as to give pleasure to my readers." His popular books such as *The Life of the Bees* and especially his autobiography, *A Biologist Remembers*, convey something of his philosophy as well as his enthusiasm and love of the natural world. Much of von Frisch's early work concerned the vision and hearing of fish, and he considered minnows as much his pet subjects as honey bees. A good biologist, he said, should regard the objects of his research as his personal friends.

The layman may wonder why a biologist is content to devote fifty years of his life to the study of minnows [he wrote]. The answer to any such question must be that every single species of the animal kingdom challenges us with all, or nearly all, the mysteries of life. . . . By deliberately restricting ourselves to very few carefully chosen experimental animals we end up by knowing our subjects very well indeed. Without this familiarity we should not be in a position to interpret their behavior under experimental conditions—just as the family doctor of old was able to notice and diagnose the slightest ailments of his patients because he knew them so well.

As a youth, von Frisch collected beetles and butterflies and kept a small zoo. The family summer home at Brunnwinkl, in the Austrian Alps, became the site of a small museum. Later it became the site of his extensive studies of honey bees, which began in 1912, when he was twenty-six, and continued, with occasional interruptions, for more than half a century. He began by demonstrating that bees (and fish) are not colorblind, as had been maintained. This led him to studies of bee communication and eventually to the discovery of their now-famous "dance language."

The discovery that bees had "words" for distance and direction understandably was received with much skepticism. Von Frisch prepared a series of meticulously detailed lectures, illustrated with motion pictures, which he presented at many universities in Europe and the United States. I heard him speak at Cornell University in 1949, soon after the major aspects of bee communication had been clarified, and joined many others in rushing out to the local hives to see for myself. Over the years he had many critics and was forced to defend his findings vigorously, sometimes devising more refined experiments to verify specific details. But he has also had staunch defenders, some of whom have added materially to knowledge of communication and navigation in the honey bee and other insects. For, of course, much of what was learned about bees applied to other insects, or at least served as a challenge to find out if it did apply Do other insects see colors the same way as honey bees; do they communicate sources of food; do they navigate via polarized light when the sun is obscured? Von Frisch lived long enough to see many of these problems solved as spin-off from his research. He died in 1982 at the age of ninety-six.

Karl von Frisch's career had many setbacks, as most careers do. When, as a young lecturer in Munich, he became depressed over the value of his research, his mother wrote him as follows:

Any work you do, steadily and devotedly, will be for humanity in the end. It is, of course, hardly ever possible to say in advance whether anything is likely to come out of such work as yours, or what it may lead to, but each step forward towards a better understanding of nature may give rise to unthought-of new discoveries.

With such a mother's advice, it is not surprising that he persevered and proved abundantly the wisdom of her words. As he readily admitted, his doubts arose from the fact that his motivation was not utility but simple curiosity. Curiosity may have "killed the cat," but it has nourished every good scientist. So, too, has the joy of discovery. In *A Biologist Remembers*, von Frisch tells how he discovered how far bees responding to the "dance" will forage from the hive. He kept watch with a dish of honey at a meadow a kilometer from the hive. His brother Hans was stationed on a hill half way to the hive, and another worker still closer to the hive. Karl was to signal to his brother by means of an old cow horn if a marked bee appeared; Hans would pass on the signal by ringing a cow bell, and the third man by blowing a trumpet, whereupon observers at the hive would watch for the return of the bee. When a marked bee arrived, von Frisch remarks: "never before or since have I blown into a cow horn with such fervour." Seldom has a scientific discovery been announced with bells, horns, and trumpets.

Alas, much modern research involves white-coated people turning dials and pounding keys in laboratories, often without even a window to the outside world. But I suspect that laboratory-bound investigators enjoy themselves as much as anyone, and are happy that they can work day in and day out without concern for rain, snow, wind, or dirty fingernails. I suspect, too, that the step was inevitable. Not that we have no further need for those who collect and classify, or for those who study the ways of animals in their natural environment. We have entered a third phase of biological investigation, superimposed on the other two: the controlled experiment, performed by a team with the aid of instrumentation never dreamed of by Réaumur and answering questions undreamed of by anyone just a few years ago.

The animals, too, have moved to the laboratory, resulting in a spate of regulations (thoroughly justified) concerning animal care. Fortunately for the entomologists, these regulations are not concerned with cockroaches, fruit flies, and the like; but one wonders if someday society's concern will also extend to these so-called "lower animals." It was the fruit fly Drosophila that first came to be an important laboratory insect, and many a geneticist has ridden to fame on the back of that one-eighth-inch-long bundle of possibilities.

The story of Drosophila began about 1900, when a Harvard graduate student, C. W. Woodworth, began breeding these flies for his embryological research. At first he bred them on Concord grapes, but when these gave out he switched to bananas, which are available all through the winter. He and his major professor, W. E. Castle, began to publish their results in 1906. In the meantime, Nettie Stevens at Bryn Mawr and Frank Lutz at Cold Spring Harbor (Carnegie Institute) had begun to use Drosophila. At Columbia, Thomas Hunt Morgan had been rearing mice, pigeons, and chickens for his studies of heredity. But these are messy animals and they take up space and eat lots of food. Morgan had heard of the success of Woodworth, Stevens, and Lutz, and he obtained a stock of Drosophila from Lutz. As Morgan wrote in a letter to a friend, there were "no funds for rearing larger animals at Columbia, [so] I got some [Drosophila] to see if I could find characters suitable for genetic work, which turned out as you know to be the case."

Incidentally, the careers of Woodworth and Lutz by no means ended there. Woodworth had a versatile career as an ecologist, physicist, and inventor, but especially as an entomologist. He became head of the Division of Entomology at the University of California at Berkeley and author in 1913 of the *Guide to California Insects*, which stimulated many a budding entomologist in that state. Frank Lutz became head of entomology at the American Museum of Natural History in New York. His *Field Guide to the Insects* became a bible for many an eastern insect collector, including me.

As for Morgan, he soon gathered around him a group of students and associates, several of whom became nearly as well known as he was: C. B. Bridges, A. H. Sturtevant, H. J. Muller, Theodosius Dobzhansky, and others. All were crowded into a small room, 16 by 23 feet, on the top floor of Columbia's biology building. Affectionately called the "fly room," the place was a glorious clutter. As described by Tove Mohr, the wife of one of Morgan's associates:

> *We . . . found it rather untidy, dusty, with milk bottles containing banana flies everywhere on tables and shelves. There was a lively conversation going on, talking, joking, laughing—the professor in the midst of the group. They were like a bunch of students having a good time together.*

Morgan was noted for his informality and his sense of humor, but he was nevertheless well disciplined and intensely absorbed in his research. He knew his own shortcomings and treated his students as colleagues, well aware that every good student surpasses his professor in some respects. One of T. H. Morgan's uncles had been General John Hunt Morgan, who led "Morgan's Raiders" on the Confederate side in the Civil War. Soon the epithet came to be applied to the occupants of the fly room. This was team work at its very best—a far cry from Fabre's lonely vigil in the fields of Provence!

Morgan received the Nobel Prize in physiology and medicine in 1933, his former student H. J. Muller in 1947. As for Drosophila, it has the distinction of having consumed more printer's ink than any other insect, with the possible exception of the honey bee. Even today, more than eighty years after its advent as an experimental animal, it still fills many a page in scientific journals. It has taught us much of what we know of genetics, and a good deal about population biology, animal behavior, and the processes of evolution. Drosophila is hereby awarded the Evans Prize for contributions to science. It carries no $50,000 stipend; only a load of bananas.

Insects, like Drosophila, that thrive on simple and easily obtainable foods tend to be favorites of experimental biologists.

Vincent Dethier, formerly at Princeton and the University of Pennsylvania, but now at the University of Massachusetts, found the blow fly ideal for studying the physiology and behavior of feeding. Dethier is a superb communicator, and he provided a fascinating account of his research in his book *The Hungry Fly* as well as a brief popular account in his delightful *To Know A Fly*.

Blow flies prefer to lay their eggs on raw liver, which is readily available and not overly expensive. In fact, says Dethier, it is such a minor item in one's research funds that "it costs more to process a purchase order through the proper channels than the liver itself costs. In truth, it is easier to purchase supplies for an elephant than for a fly." But elephants take a bit of space, and it takes twenty years to produce a new generation. Of course, a laboratory full of rotting liver is not likely to provide the best atmosphere for scientific thought. Fortunately, the flies do well on a mixture of powdered milk, brewer's yeast, and agar.

In recent years, artificial diets have been developed for many insects, permitting laboratory researchers to pick whatever species seems best for the problems at hand. Even many fussy plant-feeders can be reared in the laboratory, provided one supplies just the right combination of nutrients, vitamins, and feeding stimulants. "Cook books" for insects have been published, and several artificial diets are being produced commercially. Add to this the development of successful microsurgery on insects; the use of electron microscopes for studying ultrastructure; and the use of techniques for analyzing minute quantities of chemicals produced in their many internal and external glands. William Kirby was surely right in calling insects nature's "valued miniatures [to which] she has given the most delicate touch and highest finish of her pencil." If only he could step into a modern laboratory!

Many of the persons involved in the blossoming of modern entomology—von Frisch and Morgan, surely, and many others, would in fact not call themselves entomologists, but biologists. They are concerned with life processes, with the ways that organisms meet the contingencies of life, with the way structures and behaviors have evolved. Insects are often the most suitable organisms. They breed rapidly, they take little space, they are cheap to maintain, they are immune to animal care regulations. If a scientist can keep animals in a desk drawer, and throw them a crumb now and then, so much the better! There is something significant to be learned about every one of the million or so species. Von Frisch did not foresee the remarkable "dance language" of the honey bee, or Morgan the many problems that Drosophila would eventually illumine.

I prefer to define entomology in this way: as the study of insects as one of the best ways to revel in the joys of discovery. So I embrace every biologist who studies insects, whether he likes it or not. There are other images of the entomologist: as one who tells you when to spray your cabbages, for example. Entomologists, after all, do have to earn their keep, either by teaching, by doing research somehow relating to human society, or by advising farmers and others. No one will pay them for enjoying themselves. The day of the amateur, defined as one who has other means of support and studies insects just for sport, is largely past. The day of the amateur, defined as one who loves his work, will not pass (the word amateur is, after all, derived from the Latin *amare*, to love). Insects are too challenging, too much fun for that ever to happen.

2

The Lovebug

Lovebugs spend as long as three days—most of their adult lives—in continuous copulation. It is easy to see why they are called lovebugs, even though they are not really bugs (a term more properly applied to insects with sucking beaks, such as bed bugs and stink bugs). They are flies, rather like bulky, nonbiting mosquitoes ("honeymoon flies" is an alternate, slightly less inappropriate name for them). And to speak of love is to endow them with sentiments they are presumably incapable of.

Sex, in insects, is as much a matter of irresistible urges as in any animal, but one supposes without the psychic hangups we experience. Do insects enjoy sex? They surely do, in the sense that they are fulfilling the major goal of their existence—to make more of their kind. But it seems doubtful that their tiny brains have pleasure centers that can be activated by sex, or by drugs or electrical stimulation, as ours can. In any case, humans, even entomologists, find no erotic pleasure in watching insects copulate. A pornography of insects would never make the bestseller list, even though insects have invented every conceivable way of mating, all the way from remote control to devouring one's mate in the process.

Lovebugs would have passed as unnoticed as most small insects if they did not sometimes become annoying to the point of being intolerable. They mate in vast swarms, often in populated areas, and attract a lot of attention as they drift about with their genitals interlocked. Although lovebugs belong to a widely distributed group of flies sometimes called "March flies" (even though one rarely sees them as early as March), the lovebug proper (though scarcely proper in another sense) is a subtropical insect, ranging up the east coast of Mexico to Louisiana and to Florida. In these climes it is little appreciated by vacationers or by the natives, and so a good deal of effort has gone into investigations of its curious sex life. In the words of L. A. Hetrick, of the University of Florida, with reference to the great numbers of lovebugs that swarm in that state twice a year:

> *Adult flies are a nuisance when they spatter on automobile windshields at usual highway speeds. Driver vision is impaired and filling station attendants expend considerable time and effort removing the spattered eggs and fly remnants from the glass of windshields and headlights, and the fronts of vehicles. Large numbers of flies drawn into the cooling systems of liquid-cooled engines may cause overheating of motors, resulting in extensive damage. . . . The flies drift into freshly painted surfaces, and exterior painting of buildings is often suspended in May and September.*

Lovebugs develop from grayish, wormlike larvae that occur in damp places where there is decaying vegetation, such as rotting plant material, grass clippings, fallen Spanish moss, or cattle dung. They feed on these materials and probably play a useful role in reducing them to soil. After several weeks as larvae, they form pupae in the soil. Then, over a short span of days, they emerge in great numbers over their breeding places and drift over meadows, gardens, and highways. They have been seen as high as 1,500 feet in the air, and fishermen have reported flights of lovebugs well off the coast. For short periods, some years, lovebugs besmirch in a very literal sense the image of Florida as a semitropical paradise.

After their prolonged period of mating, the females break away from the males and fly to suitable breeding sites. Here they lay several hundred eggs, which hatch in a few days to produce the unprepossessing and generally unnoticed larvae.

Lovebugs (actually flies) spend most of their adult lives drifting about in a tail-to-tail embrace.

P. EADES

Close up, viewed with a microscope or hand lens, an adult lovebug is a far from endearing creature. It is a hunchback, the hump giving rise to a pair of fragile wings and having suspended from it six rather bristly legs. In front is the head and behind, the slender abdomen, which terminates in complex genitals, to be described below. The head is especially curious, with a short proboscis used for lapping nectar from flowers and a pair of eyes that are larger in the male than in the female. Each of the male's eyes can be seen on close inspection to be sharply divided into an upper and a lower part. In the lower part, the facets (the individual lenses that make up the compound eye of insects) are of rather normal size, and presumably serve the insect in such routine matters as landing and feeding. In the larger, upper part of the eye, the facets are much larger.

Lovebugs are not the only insects with eyes of this kind; they also occur in black flies and even in unrelated insects such as mayflies, which have similar mating swarms. Evidently these large facets have better resolving power for flying objects than the small, lower facets, which are superior for resolving stationary objects. Just why this should be so is not fully understood. The study of flies' eyes is not a top priority item in a society concerned with its possibly imminent nuclear extinction.

Swarms of lovebugs would seem to provide a feast for swallows, flycatchers, and other insect-eating birds. But evidently predators hardly ever touch them. People who have tasted lovebugs report them to be terribly bitter. Their distastefulness probably explains the significance of their unusual coloration (for a fly): the middle part of the body is bright orange. Orange is the common "warning color" of insects—as seen in other distasteful species such as monarch butterflies and lady beetles. Birds, which have excellent color vision, evidently quickly learn to associate orange with unpleasant properties. Lovebugs' orange coloration helps them not at all in avoiding being smashed in great numbers by motor vehicles.

Why do lovebugs tend to gather on highways, especially at intersections, and at filling stations, and on fresh paint and warm asphalt? Two Florida entomologists, Philip Callahan and Harold Denmark, exposed lovebugs to automobile exhaust fumes and found that they appeared disturbed, but showed no tendency to be attracted to the fumes. Nor were they attracted to methane or carbon dioxide. But when a flood lamp was turned on, lovebugs exposed to exhaust fumes flew toward the light—even though they do not ordinarily collect at sources of light. The conclusion was that light, especially ultraviolet, acts on the hydrocarbons in exhaust fumes to produce a biochemical smog which somehow mimics the odors produced by decaying organic matter in nature, that is, the sites the flies seek out to lay their eggs. As the two entomologists put it, natural sites for laying eggs cannot compete "with a duplicate attraction generated by millions of horsepower from automobiles. Well-traveled highways are saturated with invisible vapor trails that are irradiated by the UV that passes through the atmosphere." In the smog, the ultraviolet is said to act to release energy from the hydrocarbon molecules. Flights do appear to occur at times of day when the ultraviolet radiation is high, but it is only fair to say that these ideas on the attraction of lovebugs to UV-activated hydrocarbons have been met with skepticism in some quarters.

Swarming normally begins over the breeding sites. Randy Thornhill, then at the University of Florida but now at the University of New Mexico, placed dishes of newly emerged males in a pasture and dusted them with fluorescent pigment. He found that the flies hovered over the dishes for an hour or two before dispersing. In nature, swarms of males are often very dense. Dr. Thornhill collected 208 in one sweep of an insect net. The flies hover one to three feet above the ground, facing into the wind, and dart after any small object that flies near them. Females do not hover in the swarms but crawl up on vegetation and fly through the swarm, where the males compete vigorously for their favors. Sometimes several, as many as ten, grab onto a single female, each trying to mate with her. The result is a wrestling match, from which one male finally emerges with the female. Quite often one male dislodges another that has started to mate; almost always, it is a case of a heavier male dislodging one of lesser weight.

Randy Thornhill found that swarms of males are distinctly stratified, with a band of large males at the bottom, medium-sized males in the middle, and the smallest males at the top. Thus females, arising from the ground, are much more likely to encounter a large male. How a male's position in the swarm is determined is not fully understood. Males bump each other in the swarm and presumably jockey for a position near the bottom, where their chances of quickly grasping a virgin female are greatest. The female's behavior, too, is especially effective in achieving rapid coupling with a larger and presumably "better" male.

When male and female make contact, they descend to the ground or a plant. At first the male is mounted on the female's back, but after attaching firmly he turns around and faces backward. This causes his genitals to rotate 180 degrees from their original position, a seemingly traumatic movement that is, however, common to many kinds of flies. When they take flight, the slightly larger female leads, pulling the male behind her—rather like the Pushmi-Pullyu of the Dr. Doolittle stories. The male also beats his wings, and researchers at the University of Florida and the U.S. Department of Agriculture showed that single flies flew at the rate of 44 meters per minute (at 82 degrees Fahrenheit), while coupled flies achieved a speed of 51 meters per minute. This is the sort of esoteric information that entomologists love to come up with. And who is to say it is not just as important as the price of gold in Zurich on the 16th of February 1980—a figure broadcast throughout the world?

Copulating females sooner or later fly to flowers, where they feed on nectar. The male, dangling behind, may or may not be able to reach the florets. What is the advantage to the male of this prolonged copulation, when he is literally dragged about by his mate and often deprived of food? It has been shown that most or all of his sperm have been transferred within twelve hours; so why stay coupled for two or three days? The answer probably lies in the fact that when female insects mate more than once, it is usually the sperm from the last mating that fertilize all the eggs—a phenomenon called "sperm precedence." So by remaining coupled for most of his life, the male lovebug avoids competition of his sperms with those of another male. Prolonged copulation is thus perhaps properly considered an anticuckolding device.

Male and female require only an average of 4.5 minutes to achieve coupling; yet so firm is the attachment that it lasts through prolonged flight, varying temperatures, wind and rain. As might be expected, the coupling mechanism is complicated. There are three interlocking sets of valves and claspers, all of them supplied with muscles, which lock the male genitals to the female vulva. The male then discharges a packet of sperm into the vulva, and from the packet several filaments pass deeply into the female and discharge the sperm. When the flies eventually break apart, the empty sperm packet remains in the male and is soon discarded, though some of the filaments remain in the female. Both males and females are capable of mating a second time, but how often they do so in nature is not known.

Large males clearly are most successful in obtaining mates, since they attain the prime positions near the bottom of the swarm and can also dislodge smaller males that have started to mate. Collections of female lovebugs made by Craig Hieber and James Cohen of the University of Florida showed that females in copulating pairs averaged appreciably larger than nonmating virgins. Evidently males preferentially mate with larger and heavier females. Such females have a greater egg-laying capacity and so are able to increase the representation of the male's genes in the next generation. It is also adaptive for the female to have coupled with a large male, which can better assist her in flying to a good site to lay her eggs. Also, a small male may not survive the long mating flight. Dead males have occasionally been seen dangling from females, exhausted from the long ordeal. Such females are unable to disengage and so cannot lay their eggs.

So both males and females tend to prefer larger mates, for good reasons. Small size is clearly disadvantageous, but it is not the result of genetic factors; some larvae simply fail to obtain as much nourishment as others, and so produce smaller adults. If it were genetic, perhaps natural selection would produce flies of gradually increasing size. Lovebugs the size of turkey vultures would likely prove an embarrassment to the tourist industry.

Small though they are, lovebugs continue to plague Florida periodically. Why they came to the forefront only recently is not clear—the species was not even known to science until 1940. One writer speaks of the plague as an "ecological snafu," implying that the increase in tourism and cattle raising has provided many more breeding sites. The draining of swamps, the development of pastures filled with cattle droppings, and the general increase in organic wastes of all kinds, it is claimed, provide all kinds of opportunities for lovebugs. This may be true, but for some reason the insects do not attract the headlines they once did, and the state has ceased to provide funds to study them. Perhaps people are learning to admire them and to envy their lives of perpetual sex. But I doubt it. Floridians are too absorbed in the luxuries of their condominiums, too awed by the creations of Disney Enterprises, and too frightened by social changes in their cities to be much concerned with lowly "bugs." When they are surfeited with "human progress" the lovebugs will still be there, living their elemental lives in the sunshine.

3

The Flea

Once upon a time fleas were as much a part of human life as dandruff and hangnails, though doubtless a bit more worrisome. The word "flea" is said to come from the Old English flēon, to flee, a suggestion that these insects are regarded as less than welcome. And while we are at the dictionary: fleabane is a plant named for its presumed ability to drive away fleas (but fleawort is supposed to have seeds that look like fleas).

In the days when the English language was emerging from its Latin and Germanic origins, fleas were obviously on people's minds. In illustrated documents of the Middle Ages, and even into the eighteenth century, people were often shown wearing elaborate devices around their necks, not merely ornamental but in fact early varieties of "flea collars," relying on sticky surfaces rather than subtle chemicals to overcome their fleas.

In merrie England of the seventeenth century, fleas were very much a part of life, and John Donne lamented to his lady love (quoted here only in part):

Mark but this flea, and mark in this,
How little that which thou deny'st me is;
It sucked me first, and now sucks thee,
And in this flea, our two bloods mingled be;
Thou know'st that this cannot be said
A sin, nor shame, nor loss of maidenhead,
Yet this enjoys before it woo,

And pampered swells with one blood made of two,
And this, alas, is more than we would do.

Fleas appear many times in literature and song. In Auer-
bach's cellar in Leipzig, Mephistopheles sings (according to
Goethe):

A king there was once reigning,
Who had a goodly flea,
Him loved he without feigning,
As his own son were he!

The king, so the ballad goes, dressed his flea in satin and
velvet and made him a prime minister; whereupon the queen
and her attendants were bitten but dared not scratch themselves.

Fleas are only moderately discriminating in their tastes.
Each of the about 1,500 species of fleas tends to have a favored
host mammal or bird, but will in many cases take advantage of
any convenient source of blood. There is a so-called human flea,
suitably called *Pulex irritans* (the Romans thought fleas arose from
dust, and the word "pulex" is said to be derived from *pulvis*, the
word for dust). But the human flea is believed to have originally
been a pest of pigs. (The human body louse can survive only on
people and pigs; do these insects sense some subtle kinship?)
People are more often bitten by cat, dog, rat, or chicken fleas.
But, of course, in our sanitized Western world it is uncommon to
be bitten by a flea at all.

It was not always so, and in parts of the world today being
flea-bitten is not all that uncommon. It may be no coincidence
that about a fourth of all the research papers published on fleas
emanate from Russia; or that Modest Moussorgsky wrote *two*
"Songs of the Flea." Fortunately, the fleas that infest wild animals flourish everywhere, so we may still enjoy those marvelously adapted creatures if we care to dig into the nest of a rabbit
or a bank swallow—or to adopt a stray cat or dog.

Fleas attack only warm-blooded animals. Evidently mammals were their original hosts, as only about a hundred species
occur on birds, and these are a diverse group evidently derived

secondarily from groups occurring on mammals. Oddly, none attack single-hoofed mammals (horses, tapirs, and the like) and none attack monkeys and apes except casually. None attack aquatic mammals, yet many sea birds have fleas, even Antarctic penguins. Mammals have been on earth for something like 180 million years, birds for a shorter time than that. Insects have been around for perhaps 300 million years. Presumably fleas evolved soon after the mammals appeared, perhaps from some kind of fly that scavenged in the nests.

Fleas are tough-bodied creatures, as anyone can confirm who has tried to crush one in his fingers. Their bodies are so flattened from side to side that one wonders where there is space for all their internal organs. The compressed body is, of course, marvelously suited for slipping between the hairs or feathers of their hosts in their search for blood. Perhaps the most striking feature of fleas is the presence of rows of thick spines, forming "combs," chiefly on the head and just behind the head. Each species of flea has its own particular arrangement of these combs, evidently depending upon the kind of animal it normally attacks. Often the distance between the comb-spines is closely correlated with the diameter and density of the hairs of the host animal. Fleas that attack bats have especially strong combs, enabling them to cling to their hosts while they are in flight. The porcupine flea is said to have the largest combs of all. But some fleas have no combs, chiefly those that bury themselves in the skin and cannot be easily removed. Scratching, grooming, and preening are the major defenses of mammals and birds against fleas, lice, and other parasites, and the combs and claws of these insects play a major role in preventing dislodgment.

The jumping abilities of fleas also invite our admiration. Fleas are reported to be capable of a broad jump of thirteen inches and a high jump of eight inches (but one olympian cat flea is said to have made a high jump of twelve inches)! It has been pointed out that a man with comparable jumping abilities could set a record for the high jump of 800 feet. But that is not a fair statement. All insects appear to have prodigious strength, a result of the fact that as any animal increases in size, its weight increases at a far more rapid rate than its size and strength. The numerous short muscles of insects, attached to an external skeleton, also render them capable of seemingly great feats of strength. The hind legs of fleas are unusually large and strongly musculated. After all, fleas have a limited repertory of behavior; it is not surprising that what they do they do well. Finding a host animal, or leaping from one to another, is of paramount importance in their lives. Sand martin fleas spend the winter in cocoons in abandoned martin nests in the ground, and when the martins return in the spring, the newly emerged fleas leap upon them even before they land. It is reported that if an artificial bird is held in front of the burrow with its wings flapping, the fleas will leap upon it.

Not only do fleas excel in the art of jumping, but they are able to jump repeatedly. An oriental rat flea, if stimulated by the presence of other fleas, will jump an average of once a second for as long as three days! The muscles of the hind legs show little evidence of fatigue, and the leg movements are made much more automatic by the presence of a pocket of resilin at the base of the leg. Resilin is a natural rubber capable of storing and releasing energy by contracting and stretching, producing a rebound somewhat like that of a good rubber ball. When a flea is about to jump, its hind legs are "cocked" by pulling them up so that the major joint is against a ridge on the body. This compresses the resilin. When the leg is suddenly extended, it snaps from the catch and at the same time permits the resilin to expand, providing the extra bounce.

One of the advantages of using energy stored in an elastic structure is that it is relatively independent of temperature. Insects, in general, move slowly when it is cold, but fleas that can move at all can take advantage of elastic bounce. Bird fleas have been seen leaping about on snow in the Alps. Rabbit fleas can be frozen at about 32 degrees Fahrenheit for several months, and when returned to normal temperatures will jump about within a few minutes, prepared to gorge themselves as soon as they can find a source of blood.

Fleas locate a food source by leaping about until they find a

---

Rabbit fleas remain in rabbits' ears for long periods, ready to respond to changes in the hormone levels of their hosts.

P. EADES

cue telling them a host is near. The cue may be a specific odor or an increased concentration of carbon dioxide. Evidently they also respond to vibrations. Fleas occurring in the nests of certain rodents respond to the tread of a man and may actually pursue him for some distance. Since fleas are generally able to live for weeks apart from a host and without food, they can build up a considerable appetite. That is why persons entering houses that have been empty for some time are sometimes bitten viciously; the former owners of the house had pets, and their fleas have been waiting patiently for the new owners. Starved fleas have been known to live for over a year.

Fleas are remarkably efficient at finding a host. One researcher released 270 marked rabbit fleas in a meadow of 2,000 square yards, and within a few days nearly half of them had found a rabbit. Another researcher collected 48,996 fleas from 143 ground squirrels, and within three days most of the squirrels had been reinfested with about equal numbers of fleas.

While many fleas jump off after they have filled up with blood, some have a much more intimate association with their host. Rabbit fleas remain attached to the ears for long periods of time, their mouthparts imbedded in the flesh. But curiously, they are unable to breed, and so long as they remain on a male they can never breed. When rabbits mate, their fleas become excited; at this time the temperature of the rabbits' ears rises by several degrees, and the fleas begin to hop about actively and may move from male to female. About ten days before the female is to give birth, there is a change in the hormones circulating in her blood, and this stimulates the development of eggs in the female fleas that are attached to her. This has been shown experimentally by injecting female hormones into male rabbits, causing their fleas to initiate egg development.

When the young rabbits are born, the fleas move from mother to nestlings, where they feed voraciously, mate, and lay eggs. It is believed that growth hormones in the blood of the new-born young stimulate copulation, since fleas on adult rabbits never attempt to copulate. This intimate dependence upon hormones of the host has been demonstrated mainly in the rabbit flea, but it may be more widespread than now realized. It has often been stated that women are more apt to be bitten by fleas than men. Perhaps there is truth in this, and good reasons for it.

Fleas will bite over and over again if given a chance. Their mouths are supplied with swordlike blades that lacerate the skin. An anticoagulant is pumped into the wound, and the blood is drawn into the digestive tract by a muscular pump. Blood is imbibed rapidly, and much of it passes through the body and dribbles from the anus. This apparently sloppy manner of feeding reflects the fact that the fleas extract certain nutrients that occur in blood only in small quantities, ridding themselves of the surplus, which falls into the nest where the fleas' larvae live. The rather wormlike larvae feed on scraps in the nest, including dribblings from the adult fleas. The adults in a sense "feed their babies" with fecal matter that is mostly blood. Eventually the larvae spin silken cocoons in which they transform to adults. When they emerge, they must first jump onto a suitable host animal, where they feed and may ride about for some time before receiving the appropriate cues that induce mating.

When fleas are "in the mating mood," the males release their short antennae from the grooves in which they are usually held and hold them erect. Then they slip beneath the female and use their antennae to grasp her from below. The sex organs of the male are so complex that for many years no one could figure out exactly how they worked. Some of the details were discovered by quick-freezing copulating fleas and then dissecting them and studying them under high magnification. Among the spines and hooks that serve to grasp the female is a delicate structure looking somewhat like a feather duster. Evidently this serves to titillate the female; or perhaps it serves to detect a subtle odor emanating from her. The penis is a large structure, often half as long as the male's body, but coiled like a watch spring within it. When erected, the major part serves as a guide for a shaft that penetrates as far as the female's genital pouch, where it locks in place and itself serves as a guide for a slender, flexible rod that extends to the sperm receptacle, deep inside her abdomen. Here the sperm are discharged, and stored until released to fertilize the eggs as they are laid.

It is difficult to understand how such a remarkable structure as the flea's penis may have evolved. Perhaps the concept of sperm precedence, mentioned with reference to the lovebug, provides a clue. It may be important to the male not only to deposit sperm directly into the sperm receptacle, but to place it in a position where it will be most likely to fertilize the female's eggs. Presumably this would be near the receptacle's entrance, thrusting aside any sperm from earlier matings. (Some male damselflies are even equipped to scoop out the sperm from previous matings before depositing their own!)

Many of the intimacies of fleas' lives have been exposed and elegantly described by the world's leading flea enthusiast, Miriam Rothschild. Miss Rothschild was literally brought up on fleas—many dead ones, on microscope slides—since her father was a noted authority on the classification of fleas. Charles Rothschild was a successful banker in the Rothschild tradition and author of a book on international finance. "But what I would really like," he was heard to say, "is to be a professional bug-hunter." Charles's eccentric older brother Walter, the second Lord Rothschild, was also hooked on natural history. He is said to have taken a flock of kiwis with him when he went to Cambridge University, and to have had a team of zebras that he occasionally drove down Piccadilly.

Walter established a natural history museum in two cottages at the family estate at Tring, in Buckinghamshire, and selected as curator a brilliant young German biologist, Karl Jordan. When not busy at the bank, Charles spent many hours with Jordan, who nurtured his interest in fleas and meticulously illustrated his publications. He traveled widely and amassed a huge collection of his favorite animals. In 1902 the London *Express* ran a feature article on his lifelong passion, with a banner headline on the front page: "10,000 Fleas Mr. Rothschild's Hobby" (right below the headline "Mr. Kensit Stabbed" and presumably only slightly less newsworthy). When he died in 1923, his collection went to the British Museum, which eventually published a many-volume catalog of fleas, edited by G. H. E. Hopkins and by Charles's daughter Miriam.

Jordan outlived Charles Rothschild by many years and himself became a world authority on fleas as well as director of an expanded Tring Museum and recipient of many zoological honors. Miriam has remained more intrigued by the living flea and has contributed greatly to knowledge of flea physiology and behavior. A brilliant writer and lecturer, she has gone well beyond the flea into animal toxins, bird behavior, and even into psychiatry.

What is there about the flea that has attracted so many good scientists (and we have by no means mentioned all of them)? What good is a flea? There is, of course, the old saying that dogs have fleas to keep them from thinking about being dogs. Flea circuses were once popular; fleas were "trained" to leap over obstacles, to pull toy carts, and the like. Of course fleas cannot be taught anything, but they can be counted on to jump, and when harnessed to something can be counted on to pull it with their seemingly prodigious strength. Perhaps flea circuses will some day come into style again, now that pet rocks and similar fads have run their course.

Fleas were put to a better use in the fight against the rabbit scourge in Australia. Myxomatosis virus was introduced into that continent in 1950 to control the myriads of rabbits that were overrunning the land. The disease is commonly transmitted by mosquitoes, but these are not usually plentiful in the vast, arid expanses of the Outback. So in 1968 rabbit fleas were introduced as well, as they also transmit the disease. Over 10,000 fleas a day were hatched in laboratories, and these were distributed at strategic points in the field. Thanks to myxomatosis and its insect vectors, rabbits are now far less plentiful in Australia than they were at one time.

But in general fleas are pests of the first order, however much we may admire them as beautifully adapted organisms. Not only do they reduce the quality of life for many animals (and I presume even a rat may lead a quality life according to its own lights), but they transmit a variety of diseases. Chief among these, from a human point of view, is bubonic plague, the notorious black death of history, which has been said to have delayed human progress by at least two centuries.

Cities of the past were often overrun by rats, but it was not until about 1900 that the Oriental rat flea was incriminated as the major vector of the plague bacillus. The ancient world was decimated by plague in the sixth century, hastening the final decline of the Roman Empire. In the fourteenth century, the black death is said to have killed one-fourth of the population of Europe. London's great plague of 1665 inspired Daniel Defoe's fictionalized *A Journal of the Plague Year* and was chronicled in detail by Samuel Pepys in his *Diary*.

> *I walked to the Tower [wrote Pepys on October 16, 1665] but Lord! how empty the streets are, and melancholy, so many poor, sick people in the streets full of sores; and so many sad stories overheard as I walk, everybody talking of this dead, and that man sick. . . . And they tell me that, in Westminster, there is never a physician and but one apothecary left, all being dead.*

There have been no major epidemics of plague in recent years, largely as a result of better sanitation in our cities and knowledge of the causative organism and its vector. But plague is still with us, and every year cases are identified in the western United States. In this instance the natural reservoir is not rats but ground squirrels, and the vectors are several species of fleas that infest those rodents. Teams of the U.S. Public Health Service monitor wild rodent populations carefully, especially in parks and camping areas, and when plague-infected rodents are found, the area is evacuated. Foci of plague occur in many parts of the world, ready to break forth in epidemic proportions if our vigilance is relaxed. Albert Camus's novel *The Plague* provides a frightening scenario of this dread disease in a modern city.

The odd fact is that plague is not a "normal" disease of either rats or man, and the rat flea is not a "good" vector. The plague bacillus infects various wild rodents, which are somewhat resistant to it; but if a rat is bitten, it often succumbs, and its fleas move on to another rat, and so forth. Rat fleas, despite their other excellent adaptations, cannot handle plague bacilli, which multiply in their digestive tracts and often clog them, so that they can no longer feed. Some fleas die of starvation, but others succeed in regurgitating the mass of bacteria while trying to bite. That is the major way that bubonic plague is transmitted, by regurgitation into wounds in the skin by fleas that have left an infected rat and have become blocked by bacteria. It is at least good to know that fleas sometimes fall short of perfection.

So much for the flea, a poorly understood and little appreciated member of the insect underworld. Its thirst for blood is sometimes its undoing, but much more often the undoing—or at least the bedevilment—of birds and mammals of many kinds, not excluding ourselves. Surely the flea deserves our reluctant admiration. It would be hard to design an organism so admirably equipped for its own unique lifestyle. There is nothing like a flea. Thank God, you may say; but since fleas there are, we may as well hold them in the esteem they deserve.

4

The Boll Weevil

In recent years, in the United States Congress, a group of conservative southern Democrats have become widely known as "boll weevils," perhaps because they so effectively punctured traditional liberal Democratic programs. Entomologists were delighted that insects had made it to the halls of Congress even though, at the same time, they found their research funds dwindling and their children losing their school lunches. Weevils are like that: they nibble away at the good things of life.

Although weevils are only one of many kinds of beetles, they are one of the most successful—some 40,000 species in the world as a whole, as many as all the species of backboned animals combined. A weevil is just a hard-shelled beetle with its mouthparts at the end of a snout (hence the alternate name, snout beetle). Evidently the snout proved the secret of success, a device for drilling holes in tough plant tissue in order to chew out the soft insides or to lay the eggs there. Weevils and their soft-bodied larvae are specialists on fruits, nuts, seeds, stems, and roots. Grain elevators and storage bins often provide them with an especially easy life, and it is probable that all of us have eaten a good many weevil fragments as food contaminants. No harm done, as long as we don't see them.

Until 1843 the boll weevil lived an obscure life in central Mexico, probably feeding on wild relatives of the cotton plant. In that year specimens found their way to a specialist in Sweden, Carl Heinrich Boheman, who gave them the name *Anthonomus grandis*, the first compounded from the Greek words "flower ulcer," presumably an allusion to the weevil's manner of chewing holes into flower buds. As for *grandis*, Latin for "big": only an entomologist would call something one-fourth-inch long big!

Fifty years after its discovery, the boll weevil made its appearance in southern Texas, and before it lay fields of cotton planted for its convenience all the way from the Rio Grande to the Roanoke. It took the weevil only thirty years to reach the Roanoke, covering an average of about sixty miles a year despite frantic efforts to block its progress. It came to a halt only when it reached the limit of cotton growing and its own ability to survive the winter.

Boll weevils spend the winter in various kinds of vegetation, debris, or buildings, sometimes at some distance from cotton fields. Not all survive, but those that do fly or crawl to young cotton plants and feed on the tender foliage and the developing flower buds (called "squares"). They make holes in the squares with their snouts and lay an egg in each hole: when the cottony fruiting bodies (bolls) begin to develop, the weevils do the same. A female may lay as many as 300 eggs in her life span—each egg usually in a different square or boll. The egg hatches in two or three days to produce a white, grublike larva that eats out the inside. In one or two weeks the larva is fully engorged and forms a pupa, from which an adult weevil emerges in only three to five days. Thus the entire cycle from egg to adult takes only two or three weeks, allowing time for several generations in the course of the summer, occasionally as many as eight or ten. Obviously, populations can increase enormously in one year if conditions are favorable. In the late summer, adult weevils feed voraciously and build up their fat reserves to hold them through the winter. At that time, they may disperse widely, sometimes flying many miles before entering hibernation.

When the boll weevil first crossed the border into Texas, its capacity to destroy the squares and bolls of cotton became all too apparent. After a preliminary study, it was decided to outlaw the growing of cotton in Texas, and so force the weevil to retreat

---

The boll weevil, though recipient of more of mankind's pesticides than any other insect, has by no means breathed its last.

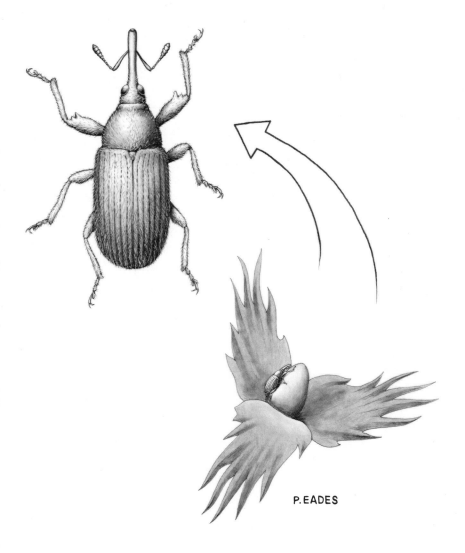

P. EADES

into Mexico. A bill to that effect introduced into the Texas legislature failed to pass, however, and the weevil moved on. In 1899 Texas appointed a state entomologist to deal with the problem, and a prize of $50,000 was offered for the discovery of a practical way of dealing with the pest. When this, too, failed, Texas asked for federal assistance.

The initial recommendation was to plant early and to plant early-maturing varieties, then plow or burn the plants after harvest. In this way a crop could be produced before the weevils had a chance to build up a large population, and their overwintering sites could be destroyed. Experimental fields handled this way had fewer weevils, but farmers were not convinced. The weevil moved on, entering Louisiana in 1904 and crossing the Mississippi in 1908, to the growing alarm of cotton growers throughout the South. In the words of L. O. Howard, chief government entomologist in Washington:

> *One who has never lived in the South cannot appreciate what this meant. At the time of the weevil's advent, so large a measure of the prosperity of that part of the country depended upon the cotton crop that its loss affected virtually every industry and every individual. The pest spread year after year and paralysis followed for a time. Mortgages on old plantations were foreclosed; negro labor fled before the weevil's advance; wealthy families were reduced to comparative poverty; banks failed; planters and speculators committed suicide.*

According to one estimate, losses to cotton in the United States in the first fifty-five years of the boll weevil invasion amounted to between four and five billion dollars. The cost of cotton goods of necessity increased at a time when cotton was facing competition from synthetic fibers. State and federal governments were strained to provide funds to learn more about the weevil and how to control it effectively. Laboratories for the study of cotton insects were built in Texas, Louisiana, and South Carolina. Labeling the weevil a "superpest," the U.S. Department of Agriculture in 1960 devoted over a million dollars to the construction of an elaborate new facility at State College, Mississippi.

In the meantime, the boll weevil was becoming part of

American folklore.

> *First time I seen the boll weevil*
> *He's settin' on the square,*
> *Next time I seen him*
> *He had his family there, . . .*
> *Tell me, boll weevil,*
> *How'd you get yo' great long bill?*
> *Way out in Texas,*
> *Among the western hills. . . .**

In New Orleans, they sang the Boll Weevil Rag, and in bars in Memphis the Boll Weevil Blues. How could such an insignificant creature have such an impact? The answer, of course, was its ability to reproduce rapidly and to disperse readily to new fields, which were abundantly available. Feeding inside the squares and bolls, it was hard for predators and parasites to reach, and equally hard to reach with chemical insecticides. Suggestions by federal and state governments that farmers convert to other crops were accepted reluctantly. John Philip Sousa had, after all, saluted "King Cotton" in one of his most inspiring marches. Was the king to abdicate? But over time other crops did replace cotton in many places. In Enterprise, Alabama, a statue to the boll weevil was erected in the town square "in profound appreciation of the boll weevil," which had brought prosperity by causing farmers to diversify their crops.

But of course cotton is still a major crop. The boll weevil has to some extent been contained, and there is even talk nowadays about its extermination. But it has been a long and expensive campaign, with a good many reverses. A mere listing and abstracting of research papers on the weevil, down to 1960, occupied 190 large-sized pages, and there has been no slowing of the pace in recent years.

Some of the early efforts to control the pest reflect the des-

---

*Sung by Vera Hall and Rich Amerson, Livingston, Alabama. Adapted by John and Alan Lomax for their *Folk Songs of North America*, copyright 1960 by Alan Lomax. Reprinted by permission of Doubleday & Company, Inc.

peration of farmers and of persons employed to advise them. Alabama entomologists recommended using a "chain drag," a log pulled by a horse or tractor between rows of cotton, with loops of link chains attached in such a way that fallen, infested bolls would be broken up and the delicate larvae exposed to the sun and killed. Another recommended device was the "loop and bag," a quick way of hand-collecting infested buds and bolls, but more effective as a way of employing lots of cheap labor than as a way of reducing weevil infestations. According to a Louisiana Agricultural Experiment Station Report of 1903, the weevils might be eradicated in newly infested fields by pulling up the plants, dipping them in kerosene, and burning them; the soil was then saturated with kerosene, burned off, plowed, harrowed, rolled, reharrowed, and rerolled. After all of that, the land was then flooded for five days! If these heroic measures did not work to check the spread of the pest, nothing would. In fact, nothing did.

Perhaps mostly to assuage the distress of farmers, the Department of Agriculture published, in 1907, a review of the importance of birds in controlling the boll weevil. Well illustrated (even with a color plate of Orioles!), the report was filled with tidbits such as the fact that of fifty stomachs of Redwinged Blackbirds examined, two contained one weevil each. In all, forty-three species of birds were found to eat boll weevils at least once in a while, while forty-three others did not seem to like them. Someone obviously had a lot of fun shooting songbirds, and some poor technician must have spent grueling hours sorting through hundreds of stomach contents looking for bits of weevils. In spite of this, it is doubtful if many farmers developed an affection for blackbirds.

An even more dubious method of controlling the boll weevil was suggested in 1904 by O. F. Cook, a government entomologist assigned to discover and import natural enemies of the pest. In a substantial technical report of the Department of Agriculture, Cook told of the discovery in Guatemalan cotton fields of the "kelep," which was said to be something like an ant but in fact was a wholly new type of insect. Keleps were reported to sting boll weevils through their armor and carry them off to their nests. Cook's report produced an outburst, in *Science* magazine, from professor William Morton Wheeler of the University of Texas, who pointed out that the kelep was in fact an ant, and a primitive one at that, and that Texas already had plenty of similar ants without importing another one. Cook's article, he said, contained "a lot of inconclusive, not to say slovenly, observations." Cook replied that "adequate ignorance of literature is a necessary qualification for learning the habits of a new insect like the kelep." This produced a further response from Wheeler, that the failure to study the available literature "may justly be interpreted as carelessness, sloth, ignorance, or conceit." So much for the kelep.

Chemical poisons were investigated early in the career of the boll weevil. But there were problems because of the vast acreages involved and the difficulty of getting at the weevil larvae inside the buds and bolls. The first was solved, beginning in the 1920s, by the widespread use of aerial application, the second by repeating application every few days so as to get the adult weevils before they laid their eggs. Fortunately, cotton is not a crop that is eaten, so one could go hog-wild with poisons as long as it was economically feasible. Through the years, cotton has been more heavily treated with insecticides than any other crop—not only for boll weevils but for several other pests as well. In recent years, about half of all insecticides used in the United States have been applied to cotton.

As early as 1898, compounds containing arsenic were being tried, and by the 1930s vast amounts of calcium arsenate were being applied to cotton fields. After World War II, arsenicals were largely abandoned in favor of the new "wonder insecticides," such as DDT. These were not only cheap, but produced spectacular results. Combined with increased fertilization and irrigation, some ten to twenty applications of the new, synthetic insecticides resulted in vastly increased yields. It was no longer considered necessary to check the fields for weevils; the insecticides were applied on a regular schedule regardless of weevil abundance. Application of DDT or a similar poison every few days was often a condition for a bank loan. Also, government subsidies were based on yield, rendering the additional expense worthwhile.

Was the boll weevil finally on the run? At first it seemed so.

But DDT and similar substances proved to be highly lethal to the natural enemies of the bollworm and the budworm (two species of cotton-infesting caterpillars), and suddenly these insects emerged as major pests—"secondary pests" in entomological terminology, since they became pests only when their natural enemies had been decimated. To make matters worse, persistent overuse of DDT and other chlorine-containing poisons induced natural selection for strains of boll weevils able to survive exposure to these insecticides—and before long bollworm and budworm had also developed resistant strains.

This problem was partially solved by switching to phosphorus-containing compounds such as methyl parathion. But bollworm and budworm gradually became resistant to these insecticides, too, resulting in the use of increased amounts, often combined with other kinds of poisons. Cotton growers, urged on by the purveyors of chemicals, became deeply infected with what has sometimes been called the "pesticide syndrome," spending so much on control of increasingly recalcitrant pests that their margins of profit declined drastically. More and more cases of poisoning to humans, domestic animals, and fish were reported, especially from methyl parathion, one of the most acutely toxic insecticides ever employed. In 1967, in the Rio Grande Valley of Texas, nearly 25 percent of the workers who loaded spray planes developed symptoms of poisoning, as did field workers and even children who were accidentally exposed. When Rachel Carson's *Silent Spring* appeared in 1962, there were many who agreed that other approaches than the mere increase in kinds and amounts of poisons would have to be thoroughly explored.

Success in the 1950s in the control of screwworm (a fly pest of livestock) by rearing and releasing great numbers of males sterilized by irradiation suggested that similar techniques might be used against the boll weevil. Fortunately, much progress had been made in developing artificial diets, based on wheat germ or cottonseed meal, enabling entomologists at the Mississippi laboratory to rear as many as 15 million boll weevils a week. Numbers of this magnitude were necessary for rearing parasites of the weevil and for obtaining the large numbers needed for sterilization and release—the idea being that females that mate with sterile males will lay infertile eggs. Unfortunately, levels of radiation sufficient to sterilize boll weevils kill them.

Chemical sterilants proved more successful, but males so sterilized often showed lessened ability to compete for females, and sometimes they regained their fertility. However, males fed a special diet for five days and then subjected to low levels of gamma radiation were found to produce normal amounts of sex pheromone (a volatile attractant important in securing a mate) and to be nearly 100 percent sterile. It now seems possible that the release of males treated in this manner soon after the weevils emerge from hibernation, when populations are low, will result in most females laying infertile eggs.

The use of this same pheromone in traps has also engendered a great deal of interest. Production of a sex attractant by male insects is a bit unusual; more commonly it is the female that produces the attractant. In this case females are attracted to males from distances of thirty feet or more downwind. Traps were first baited with live males and either coated with adhesive or equipped with a device for collecting weevils. Then, in 1969, a team of entomologists and chemists at the Mississippi laboratory announced in *Science* magazine that they had isolated the pheromone—by steam distillation of 4.5 million weevils. It was found to be a mixture of four complex organic molecules, the full names of which would require a paragraph by themselves.

It proved possible to synthesize the pheromone, and the synthetic substance is now commercially available under the name "Grandlure." The substance is so highly volatile that it disappears rapidly from traps unless evaporation is somehow retarded. One method is to impregnate cigarette filters with Grandlure, glycerin, and other substances.

Oddly, the sex pheromone, though produced by males, attracts not only females but a fair number of males. Instances of male insects tracking the signals of other males are by no means rare. Sometimes the result is an aggregation of signaling males that compete for the arriving females. In other cases some of the males "cheat on the system," intercepting and mating with females attracted to signals produced by other males. Which is the case with boll weevils is unclear; perhaps either may occur.

Traps baited with Grandlure and placed around cotton

fields provide a good way to measure boll weevil abundance, and with that information available, it is possible to apply controls as appropriate. Grandlure has also been applied to "trap crops," early plantings of plots of cotton that are treated heavily with insecticides once they have lured most of the weevils away from the main crop. While not all weevils will be destroyed in this way, the population may be reduced to a low enough point that if sterile males are released they will greatly exceed the wild, fertile males in numbers. Grandlure may not be a cure-all by itself, but there is something satisfying about turning the weevils' reproductive powers against them.

Another line of attack has been by way of the cotton plants. The early planting of short-season varieties had been suggested as early as 1900, and now we seem to have come back to the starting point. Early-maturing varieties were not formerly popular, as there was danger of damage by cold, and quality of fiber was not always the best. More recently, new varieties have been developed that are more tolerant of cold and have varying degrees of resistance to disease and to the attacks of boll weevil and other pests. Short-season varieties typically receive less water and fertilizer, and after an early harvest, the plants are shredded and plowed under before the weevils can feed to satiation before entering hibernation. The need for insecticides is greatly reduced, allowing natural enemies to restore their populations to effective levels.

So the present approach is a multifaceted one: integrated pest management, the entomologists like to call it. Populations are monitored by traps or other sampling techniques and controls designed accordingly. New varieties of cotton are being tried; insecticides and fertilizers are being used more sparingly; new techniques such as sterilization and pheromone trapping are coming into use. It is not always easy to convince farmers that these more subtle and complex approaches are better than the routine application of poisons. But according to Perry L. Adkisson and his colleagues at Texas A. and M. University, writing in *Science* in 1982, growers in the Rio Grande Valley of Texas, using the new approach, have increased their profits per acre by $31 while decreasing insecticide use from 12.3 pounds of toxicant per acre to 1.5 pounds.

A similar scenario has unfolded in the Canete Valley of Peru, where the budworm and a weevil closely resembling the boll weevil are serious pests of cotton. Here, too, the overuse of chemicals destroyed natural enemies and produced resistant strains of insects. More recently, parasites were reintroduced, planting and harvesting times were standardized so that there would be a fallow period each year, and when insecticides were needed the "old fashioned" arsenicals were employed, since insects do not readily become resistant to these. With these changes, yields in the valley became the highest in history.

With all this new knowledge, can we now wipe out the boll weevil? There are two schools of thought on that subject, and proponents of each are outspoken. A panel of the National Academy of Sciences, in 1975, favored the "containment" approach, and questioned whether the technology was available to achieve total eradication. This conclusion aroused considerable distress in the Department of Agriculture, whose members were arguing their case in Congress for massive funds to begin a program of eradication. Eventually, a compromise was struck, and more limited funds were appropriated for a trial eradication program in North Carolina. This has met with considerable success—but, some say, primarily because of successive cold winters in the Carolinas. The weevil is a hard-nosed customer, with powers of reproduction and dispersal that may well subvert all the technology we can throw at it. We shall see.

While people are talking about its demise, the boll weevil still succeeds in making the news. In May 1982, students in the Hermanos Escobar School of Agriculture in Ciudad Juarez, Mexico, took over the school and held hostage 15 million insect larvae, evidently in protest over inadequate funding for their school. The larvae were not identified, but were said by the Associated Press to be "predators which eat boll weevils threatening the area's cotton crop." Local farmers pleaded with the students not to kill the larvae, and a few days later they were released. But all had died as a result of changes in the temperature of the laboratory where they were housed.

The weevil was clearly the winner in that skirmish.

5

The Mormon Cricket

It was in July, 1847, that Brigham Young and his band of pioneers entered the valley of Great Salt Lake, then a dreary, treeless expanse—but in a vision Young had seen it filled with the "glory of Zion." Within a few days a city had been planned, houses started, and crops planted. Fortunately, the following winter was mild, and the limited harvest was supplemented by wild game, roots, and thistle tops.

In the spring of 1848, there was a prospect of a much more abundant harvest, enough to meet the needs of the many new arrivals. But in late May a vast swarm of "crickets" appeared, "black and baleful" as the locusts of the Bible. Behind them they left "not a blade or leaf," and the land was as if "scorched by fire." The settlers beat them with sticks and ropes, and swept them into piles and burned them. But for every one killed, a hundred more seemed to appear. The settlers had largely consumed last year's harvest, and they were hundreds of miles from another outpost of civilization. The situation was desperate.

Colonel Thomas Kane, one of the more prominent of the pioneers, described the crickets as "wingless, dumpy, black, swollen-headed, with bulging eyes in cases like goggles, mounted upon legs of steel wire and clock spring, and with a general personal appearance that justified the Mormons in comparing them to a cross of a spider and the buffalo." Hardly a dispassionate description, but one that reflected the feelings of the settlers!

A three-day fast and prayer were decreed. And lo, from the

lake appeared great flocks of sea gulls, white as angels. The gulls settled in the fields and gorged themselves on crickets. Each evening they returned to the lake, then reappeared at dawn to feed again. Soon the crickets were gone, and the crops recovered. In August the settlers held a festival of thanksgiving. Wrote John Young (a nephew of Brigham): "I have met those who were skeptical about the gulls being sent by Divine Providence, for the salvation of our people; but I believe in it most firmly. . . ." Some years later, the Mormons erected a statue to the gulls in Temple Square, still a major item on visitors' tours of Salt Lake City.

This is a story that has been told many times. The crickets had, of course, been around long before the Mormons arrived. In fact they formed a major (although highly seasonal) item in the diet of several tribes of Indians. Sometimes they were eaten raw, after the spiny legs had been pulled off, but more often they were roasted or put into soup. During outbreaks, they were sometimes gathered in sacks, and later dried and pounded with grass seed into a powder which could be moistened with water and baked into a cake in the winter when food was scarce.

The story of the crickets did not end with Mormons, of course. In 1871 and again in 1878 and 1891 there were serious outbreaks in Idaho, and in 1873 and in many subsequent years, outbreaks in Oregon, Washington, Colorado, and other states. In 1938 nearly 19 million acres were infested in eleven western states. In parts of northwestern Colorado, many ranchers moved away after repeated invasions of Mormon crickets had devastated their ranges. The crickets are still very much in evidence in the mountains of the West, though only occasionally do they build up large populations and migrate into cultivated valleys (by foot, since their wings are small and functionless for flight).

"Many interesting and marvelous stories have been told of the crickets," wrote entomologist Robert Milliken in 1893, "how they fill up irrigation ditches, get into the flumes of mills, stopping the wheels with the mass of their bodies, and filling rivers, forming bridges so that the advancing army can cross

---

The Mormon cricket: a female, as evidenced by the sword-shaped ovipositor.

P. EADES

over safely on the dead bodies of their comrades, etc. I can readily believe this, for when their great size is taken into account, together with the countless millions in which they occur, and the well-known propensity to go straight ahead, turning neither to the right or left, one can readily understand that an obstruction will soon take place if the advancing army keeps on its course. It is a standing joke that the crickets will not turn aside for a telegraph pole if it is in their path, but will go up one side and down the other to keep in the line of their journey."

The response to telegraph poles is doubtless a bit of an exaggeration, but many persons have reported that migrating crickets do tend to maintain a specific direction once they have set out, sometimes climbing over board fences or even buildings. In one instance a band is said to have entered an open-air pavilion in Wyoming when a dance was in progress. The dancers soon made a mash of them, making the floor so slippery that the dance had to be discontinued. How the crickets "decide" in which direction to set out remains a mystery.

The Mormon cricket, it should be explained, is not really a cricket at all. Rather it is a katydid of bizarre form. The robust body is about an inch and a quarter long. If one includes the long ovipositor (egg-laying tube) of the female, the length is well over two inches; and if one adds the long, threadlike antennae the total is some three and one-half inches. Perhaps the most notable feature is the pair of very short front wings, scarcely visible behind the large thoracic shield of the male (who uses the wings to produce a "mating song") and not visible at all in the female. The hind wings, which are the major flight wings in most katydids, are absent in both sexes. The female's ovipositor is up-curved and swordlike, while the abdomen of the male terminates in a pair of curved and pointed claspers, rather like powerful jaws at the wrong end of the animal. In color the crickets vary from black or brown to bright grass-green. No one has fully explained the significance of the color variation, but it appears that when populations are low, green colors predominate. However, when there is a population explosion and great hordes move into the valleys, most crickets are black, with greenish legs, and sometimes greenish heads. Add to their size and polyglot coloration their long spiny legs, and it is easy to see why a single one might instill fright in a person, to say nothing of a few hundred thousand of them.

A few years ago I encountered a band of the crickets high in the San Juan Mountains of Colorado, at an elevation of about 10,000 feet. That was by no means a record, as the crickets have been seen as high as 13,000 feet. Migratory bands are able to advance about a mile a day, and bands have been found covering a square mile or more, with several hundred crickets per square foot. More than 250 kinds of plants are known to be used as food, mostly meadow and range plants, but the crickets will also consume wheat, alfalfa, garden crops, and even small fruits. They also eat one another, dead or alive. Crickets that have lost a leg or otherwise become handicapped are quickly cannibalized. When bands cross a highway, they are crushed by the thousands, and others appear to feed on the remains, resulting eventually in a black and slippery mess that may cause an accident.

One of the strangest features of these strange animals was described many years ago, in 1904, by a pioneer Colorado entomologist, C. P. Gillette. He noted that many females had white masses attached to their tail end (ranchers called them "blubber"). Gillette discovered that these masses are produced by the males and are transferred to the females during copulation. "The females carry these conspicuous white objects about for a time," Gillette said, "extracting a portion, at least, of their contents for the fertilization of the ova." More of this later.*

For the moment it is worth digressing to say a word about Gillette, who did so much to increase knowledge of the then little-known insect fauna of the Rocky Mountains. He was born and educated in Michigan, but in 1891 was given an opportu-

---

*Although Gillette was the first scientist to describe these "white objects," in fact they had been noted almost half a century earlier, in 1859, by Captain John Feilner while exploring California under the auspices of the Smithsonian Institution. After the insects mate, he reported, "a small bag . . . is attached to the body of the female close to the tail; this is extracted from the other without the tail; after a while the bag disappears." These and other remarks make it clear that Captain Feilner had encountered the notorious Mormon cricket, though he did not identify it as such. The good captain was killed by Indians a few days after he made and recorded these observations.

nity to take charge of the new Department of Zoology and Entomology at Colorado Agricultural College, in Fort Collins. Later
he became Colorado's first state entomologist, still later director
of the Agricultural Experiment Station, and eventually vice-
president of the college (now Colorado State University). Well
aware of the problems in tackling such a largely unknown insect
fauna, he proceeded to build up a collection of Rocky Mountain
insects, which by 1897 already included 40,000 specimens. He
kept up an active correspondence with entomologists all over
the world and published many papers on aphids and leafhop-
pers as well as on insect problems of concern to Colorado
farmers. Unfortunately, both the insect collection and the Experiment Station have languished since Gillette retired in 1933.

   In Gillette's time, travel in the West was largely by rail, and
"the Professor" became well known to conductors as he traveled about collecting insects and meeting with farmers. It was an
outbreak of Mormon crickets not far from Steamboat Springs
that brought his attention to these insects, then more of a legend
than an object of scientific inquiry. Since Gillette's time a good
deal more has been learned. The "white masses" are indeed
transferred from male to female during mating, and she does
indeed extract material from them, but there is more to the story
than that. Credit for working out the details belongs to Darryl
Gwynne, a former graduate student at Colorado State University
(though long after Gillette's time) who finds the Mormon cricket
a model of unusual biological interest.

   In katydids, generally, males produce a characteristic song
by rubbing parts of their front wings together. The song may be
loud and resonant, as it is in the katydids of the woodlands of the
eastern United States. Or it may be thin and rasping, as it is in
Mormon crickets and many other species. Larger males, by and
large, produce louder songs and are more successful in competitive interactions with other males. Females respond to the song
by approaching a male, in some cases themselves producing an
answering chirp.

   In most insects, indeed in most animals, the female is the
"coy" sex, responding cautiously to a male whose courtship is
most vigorous—assuring a strong genetic input for her progeny.
This makes sense, for it is the female that will bear the burden of

producing the offspring, and it is to her advantage to be choosy. The male, on the other hand, produces an abundance of tiny sperm but, in most cases, has much less to do with the production and rearing of offspring. It is to his advantage to compete with other males so as to fertilize as many females as possible. In this way his genes will be abundantly represented in the next generation. But here and there one finds examples of "sex reversal," that is, cases in which females compete among themselves for males, and males are the choosy sex. Mormon crickets, it turns out, provide a striking example of sex reversal, and it is the "white masses" produced by the males that provide the explanation.

When Mormon crickets are migrating, they compete vigorously for whatever food they come across, but there is little sexual behavior and males do not produce their mating songs. But in the morning, after the swarm has roosted in low vegetation for the night, a few males will climb up into shrubs and begin to sing. It is these males that have produced "white masses" (properly called spermatophores), which in fact are so large that they make up as much as 30 percent of the males' weight. They do not usually sing for long, for females are quickly attracted and often fight among themselves for access to a male. Females attempt to mount the male and to link their genitals with those of the male. But at this point the male may reject the female and move away quickly.

Dr. Gwynne collected equal numbers of rejected females and females that were accepted by males. He found that accepted females weighed significantly more than rejected females, suggesting that males actually "weigh" the females when they mount and accept only heavier females. These "heavyweights," Gwynne found, have larger ovaries that contain more mature eggs. The males have, after all, invested a great deal of energy in the production of these huge spermatophores—equivalent, says Darryl Gwynne, to an average-sized man losing some sixty pounds at each mating! It is to his advantage to see that this effort pays off in terms of the representation of his genes in the next generation. Not that either males or females think out these matters, of course—they are products of Darwinian natural selection, in which the currency is the perpetuation of one's genes.

The spermatophores of Mormon crickets are not merely packets of sperm; they also contain a large mass of protein. After mating, the female reaches beneath her and consumes a large part of the proteinaceous material, which is rather stringy, having in Gwynne's words, "somewhat the consistency of mozzarella cheese on a warm pizza." When males are fed protein labeled with radioactive carbon, the protein can be found in the spermatophores and ultimately in the ovaries and mature eggs of females he has mated with. Evidently the females depend upon this dietary supplement and will fight among themselves for it. Males, on the other hand, do not dole it out readily to any female that comes along—she must be a good risk in terms of the currency of natural selection.

All of these remarks apply to Mormon crickets when populations are high. When thousands are moving across a meadow, there is severe competition for whatever food is available. Only a portion of the males produce spermatophores at any one time, and females strive to obtain them. But when populations are low, there is more food to go around, and all are well fed. Males are able to develop spermatophores more readily, and females have less need for protein to develop their eggs. Under these conditions the roles of the sexes are more normal—males compete among themselves for access to females, and females do the choosing. Masculinity and femininity are thus relative and sometimes ambiguous words, subject to change depending upon circumstances—at least in Mormon crickets.

That the roles of the sexes are subject to change does not seem surprising to us, living as we do in a fluid and fragmented society. A male washing dishes while his spouse competes in the marketplace no longer seems strange to us. But we think of insects as being bound by their genes to very limited regimens. Mormon crickets provide an unusual example of shifting sex roles in a creature that was once described as built of "steel wire and clock springs." Who knows what else may lie behind the dim eyes of insects?

6

The Gypsy Moth

Readers in the northeastern United States may wish to skip this chapter; they have seen enough of the gypsy moth. On the other hand, they may find it food for thought. For more than one hundred years this resilient insect has managed to subvert our vaunted technology. The story of the gypsy moth engenders humility if not humiliation.

The arrival of the gypsy moth in North America is associated with the name of Leopold Trouvelot, a French astronomer working at the Harvard Observatory. Trouvelot was concerned over the plight of the French silk industry, which was threatened by a disease of silkworms. Believing he could develop a strain of silkworm resistant to the disease, he imported gypsy moth eggs from France in 1869 in the vain hope of crossing that species with others. How they escaped from his home at 27 Myrtle Street in Medford, Massachusetts (a suburb of Boston), is uncertain. One story has it that the eggs were blown from a window by an unexpected draft. In any case, Trouvelot was aware of their escape and the possible danger it posed, as the gypsy moth was well known as a pest in Europe. However, no one noticed that the pest had become established, so Trouvelot's warnings went unheeded.

Trouvelot returned to France, and his house was sold in 1879 to William Taylor, who discovered fuzzy caterpillars swarming about the place in numbers. Over the next ten years they gradually spread into neighboring yards, and in 1889 they occurred in outbreak proportions over much of the city of Medford. Trees were stripped of their leaves in many places, and caterpillars swarmed over yards and sidewalks and even got into houses. At night, the sound of the crunching of leaves by so many caterpillars could be heard, and the fall of so many fecal pellets sounded like rain. During the outbreak, specimens were identified as belonging to the notorious gypsy moth, and the connection with Trouvelot was established. The matter was brought to the attention of the state legislature, and in 1890 a bill was passed appropriating $25,000 for the extermination of the pest. This was the first state law in the United States calling for the extermination of a pest and it established "the right . . . to enter upon the lands of any person" in execution of the act. Thus it set a precedent that has caused a good deal of irritation, and some litigation, through the years.

The gypsy moth had been described by Linnaeus, who called it *Phalaena dispar*, "phalaena" being the Greek name for a type of moth, "dispar" an adjective referring to the great difference between the male and female (as in the word disparate). The first person to call it "gypsy moth" was apparently an English butterfly enthusiast named Benjamin Wilkes, in 1742. Perhaps the brown coloration of the males suggested to him the swarthy complexion of gypsies, or perhaps it was the vagabond lives of the moths. The history of the moth in England is interesting. It was never common there, and by the late 1800s it had evidently become extinct, perhaps because it had been so energetically collected by amateur lepidopterists. Several attempts were made to reintroduce it, mostly without success. On the continent, it had ravaged forests, orchards, and shade trees for many years, all the way from France to Japan, but outbreaks were always sporadic and unpredictable. It is no less a pest there today.

The spread of the insect from its original inoculum in Medford, up to 1896, is documented in great detail by E. H. Forbush and C. H. Fernald in their book *The Gypsy Moth*, published by the Massachusetts State Board of Agriculture. This is a curious book: Forbush wrote the first half, Fernald the second half, with a fair amount of repetition and lack of organization. Fernald was a professor at Massachusetts Agricultural College in Amherst, while Forbush was an ornithologist located in Boston. They explained in their introduction that they "have labored under the disadvantage of residing in different parts of the state"—doubtless it was a long way from Amherst to Boston in 1896!

In the twenty-seven years since its introduction, the gypsy moth had spread throughout greater Boston and was causing much alarm. The reaction was to "throw the book" at the pest—in this case a fat volume of 495 pages, with 100 pages of appendices and 66 plates, three of them in color. Eighty-five years later, after the publication of innumerable bulletins, circulars, and research reports, the U.S. Department of Agriculture has come out with a 757-page volume, with no end of tables, graphs, and colored illustrations. On my bathroom scale it weighs six pounds. If gypsy moths could read, they would surely be intimidated.

The gypsy moth now occurs in all the New England states and in New York, New Jersey, Pennsylvania, and the southern parts of Ontario and Quebec; there are also pockets in Ohio, Michigan, Maryland, and Virginia. Areas of forest defoliation are being mapped via aerial photography and Landsat satellites, and traps baited with female sex attractant have picked up male gypsy moths as far away as Florida and California. In 1975 the range of the pest extended over 200,000 square miles, and there was speculation that it might eventually include the entire hardwood forest region east of the Mississippi and perhaps even the coniferous forests of the South and Far West.

Gypsy moths lay their eggs on tree trunks in masses of several hundred, covered with hairs from the body of the female. The eggs persist through the winter, and in April or May the hairy caterpillars emerge and climb into the trees. At first they are not very conspicuous, as they feed mostly at night and hide during the day. The small larvae prefer oaks, but do well on aspen, basswood, white birch, and even apple and roses. By mid-June the caterpillars are much larger and more voracious, and may have completely defoliated the tree on which they started. The more mature larvae are even less fastidious feeders, and may move to beech, plum, and even conifers such as pines and hemlocks. When populations are very high and most of the preferred trees and bushes have been defoliated, they may consume the foliage of less-favored trees such as elm, hickory, and others. Fortunately, there are some trees that contain chemicals that are repellent to gypsy moth caterpillars: ash, catalpa, dogwood, tulip poplar, and several others. Unfortunately, oaks make up a considerable proportion of the hardwood forests of the northeastern United States and are major shade trees in many places; in mixed stands, the associated trees are usually beech, maple, pine, and other trees that are almost as readily attacked.

Most hardwoods survive the loss of most or all of their leaves to the caterpillars, even if this occurs over two or more successive years, as they usually produce new foliage in late summer, after the caterpillars are gone. This is less true of pines and hemlocks, which sometimes die after a single complete defoliation. Any loss of leaves tends to weaken a tree, rendering it more susceptible to attack by fungus or by boring insects. Woodlands denuded repeatedly by gypsy moth caterpillars tend to undergo changes in composition, the less susceptible trees tending to replace the more susceptible. When the caterpillars are active, the loss of shade causes the woodlands to become unusually warm, and plants and animals that thrive in cool, shady places die or move elsewhere. There are fewer places for birds to nest, and no acorns for squirrels to gather. So there are many problems besides the mere presence of so many hairy caterpillars in the trees and on the sidewalks and buildings of Suburbia. Persons afflicted with entomophobia (fear of insects) have a particularly hard time of it; and even those blessed with entomophilia find their relationships with the insect world somewhat strained.

So why do we put up with these creatures? Just offhand, they would seem easy enough to contain. The robust, whitish females are not even able to fly, and the egg masses they produce are conspicuous and not difficult to destroy. Unfortunately, it is impossible to find or reach all the egg masses in a given area, and just a few will produce several thousand caterpillars. During an infestation, egg masses may be attached to boxes, crates, freight cars, lumber, vehicles, or nursery stock, and thus be unknowingly carried into new areas. There are several documented cases of outbreaks occurring following the importation of nursery stock; in fact in 1911 there was a second introduction from

Europe into New Jersey on spruce trees brought from Holland.

Even without human help, gypsy moths disperse readily despite the fact that the flightless females deposit their eggs near the place where they developed as larvae. The larvae, soon after they begin feeding, spin silken mats on the underside of leaves, where they spend resting periods. They are readily disturbed by wind currents, and tend to release their grip and to hang from a silken thread, so that they are easily blown from the tree. The small larvae are light in weight and are densely covered with long hairs which trap air, such that they drift about in air currents for a considerable distance. They have been reported as high as 2,000 feet above the ground and as far as twenty miles from their place of origin, although most of them usually end up on nearby trees. A larva that happens to be on a less favored food tree, or on one that is overcrowded, tends to be especially restless and is likely to drop on a thread and be wafted away. When large larvae find themselves running out of food, they too may disperse, by walking over the ground to another tree. They are not likely to go far unless they climb into an automobile or a suitcase, as they have been known to do. So with or without our help, gypsy moths do very well at the art of dispersal.

The story of efforts to control the gypsy moth is a long one, and the final chapters have yet to be written. Forbush and Fernald, in 1896, devoted literally hundreds of pages to the subject. Eggs, they reported, can be scraped from the trees and burned; they can be painted with creosote; or they can be sprayed with carbolic and nitric acid if one is willing to handle such caustic chemicals. Small caterpillars can be prevented from ascending trees by circling the trees with sticky bands. Burlap strips placed around trees will collect larger larvae beneath them during the day, where they can be crushed or "cut in two by the knife," presumably by persons intent on personally avenging the devastation to their trees. Larvae can also be destroyed by spraying them with arsenic-containing substances, and as early as 1896 a variety of tanks, nozzles, and horse-drawn vehicles had been designed for spraying orchard and shade trees.

By 1890 the infestation of gypsy moths in the Boston area had been reduced to such an extent that the state of Massachusetts discontinued control efforts. The pest had been licked; no use spending further money on the problem. But within a few years the insect was back again in outbreak proportions and spreading over other parts of New England. (The history of entomology is filled with such episodes, and it is at least partly the fault of entomologists, who like to think that insects can be categorized as "economic species," meaning that they are worth spending money on, or "of no economic importance." With such a simplistic philosophy, there is, of course, no need to divert funds to such underlying problems as: why do population sizes of many species fluctuate so markedly in time and space?)

By 1906 the gypsy moth had once again gotten out of hand, and federal assistance was sought and obtained. The Secretary of Agriculture was authorized to take all possible measures to prevent the further spread of the pest. A laboratory was established at Melrose, Massachusetts (another suburb of Boston), and it became a center of research. Improvements were made in the production and application of lead arsenate sprays, but so widespread were outbreaks that the best that could be done was to treat badly infested residential areas. Also, quarantines were established such that products likely to carry the eggs or caterpillars could not be shipped from infested areas without careful inspection. Eventually, in 1923, a barrier zone 250 miles long and 25 to 30 miles wide was established along the border between New England and New York. Within the zone, efforts were made to eradicate any gypsy moths that made their appearance. But ultimately this, too, failed.

From the beginning, much attention had been paid to birds, mammals, and native insects that attacked gypsy moth eggs and caterpillars. It became part of the federal program to import parasitic and predatory enemies of the gypsy moth from Europe and Japan. A number of promising species became established, including a large ground beetle that feeds on larvae and pupae, two kinds of flies that parasitize the larvae, and several small wasps that attack either the eggs or the larvae. By 1923, the U. S. Department of Agriculture reported, these imported natural enemies had become quite abundant, and "the results of their

---

Gypsy moth caterpillars are notorious for their appetite for foliage of many kinds.

P. EADES

work were very noticeable over much of the infested area." In 1929 it was reported that 93,000,000 enemies of the gypsy moth and the related brown-tail moth had been liberated. These belonged to forty-seven species, of which fifteen were definitely established. Unfortunately, this "shot gun" approach proved to have its problems. One small wasp turned out to be a hyperparasite—that is, it principally attacked not the gypsy moth but other parasites of that insect. Serious gypsy moth outbreaks occurred in 1924–1927, but the importation of parasites continued until 1933, when deteriorating international relations made it more difficult.

The picture changed dramatically when DDT became available after World War II. It became quickly apparent that the perfect control for gypsy moths (and many other insects) was at hand. Sprayed from planes at a rate of about one pound per acre, in an oil solution, DDT killed virtually 100 percent of the caterpillars. In 1956 nearly a million acres were sprayed, in both forested and suburban areas. Clearly there was no need for further research; DDT solved all problems.

But not everyone welcomed the rain of poisons on their private property, and innumerable lawsuits resulted. Beekeepers found that most of their bees had been killed; garden crops contained unwanted residues of DDT; and DDT was found in the milk of cows that had grazed on treated pastures. All of this was grist for the mill of Rachel Carson, whose 1962 book *Silent Spring* had a good deal to do with the gradual phasing out of DDT, beginning in 1963, and the total banning of DDT by the Environmental Protection Agency in 1972. The substitution in recent years of less persistent insecticides, especially one called Sevin, has only partially solved the problems, as these substances are lethal to bees and to many of the natural enemies of the gypsy moth. There is no doubt as to their effectiveness against gypsy moth caterpillars, and there are many persons (especially those with interests in the chemical industry) who would like to see widespread aerial spraying with these newer insecticides. We live in a remarkably health-conscious society, with increasing awareness of the dangers of drugs, cigarettes, and diverse poisons in the home, to say nothing of hazardous wastes, acid rain, and radiation escaping from nuclear power plants. In such a society it makes little sense to condone the dumping of poisons of any kind from the air on to parks, woodlands, and suburban areas.

Fortunately, the importation and release of parasites and predators has been resumed, making up some of the losses incurred during the DDT area. The sex pheromone of the female gypsy moth has been identified and synthesized, and it is available commercially under the name "Disparlure." It has been widely used to trap males, to determine their distribution and abundance, and may someday be used in large quantities to disrupt mating and cause the females to lay infertile eggs.

Gypsy moth larvae are subject to various bacterial and virus diseases, and much attention is now being devoted to the possibility of producing the germs in large quantities and spraying them onto infested trees. The disease organisms have no effect upon man or higher animals or upon the insect enemies of the gypsy moth. They do require a high degree of sophistication in production and application, and not all the problems have been solved. Wilt disease has been noted in Europe and North America ever since entomologists began to study the gypsy moth. It is now known to be caused by nucleopolyhedrosis virus or NPV, and progress has been made in producing it in quantities and applying it to forested areas. NPV has been fed in large doses to dogs, rats, honey bees, and a variety of other animals, all without noticeable ill effects. In 1978 it was registered by the EPA for use as a control agent. Wilt disease is present in nature at all times, but is most apparent following peak outbreaks of the gypsy moth. If it can be disseminated in appropriate areas prior to such outbreaks, it may be possible to block their development.

Unfortunately, all these biological methods of control require a good deal of patience and much careful research, and may result in less than perfect control. There are still many who long for the "good old days" of aerial application of poisons. In his book *Man's Plague?*, Vincent Dethier cites a report in the *New York Times* of November 11, 1973:

> *The Pocono Mountains Vacation Bureau, representing scores of resorts in this region, has started a campaign to appeal to Congress to*

*have the Federal ban on DDT lifted temporarily to permit the spraying
of area forests. The agency's president, Paul Asure, contends that
'DDT is the only hope left' in combating the gypsy moth. . . . He said
that further destruction of the forests could threaten the future of the
area as a resort center.*

In 1981 gypsy moth caterpillars stripped nine to ten million
acres of forest. Living in an area overrun by the pests, said one
official of the Department of Agriculture, was like "living in an
Alfred Hitchcock movie." *Time* reported estimates of 30,000 larvae per tree, each eating five to ten leaves a day. As that magazine described the infestation:

*Squishy little creatures hanging from branches, crawling over the
sides of houses, creating a sickly goo on roads. People rushing to their
doctors with rashes on hands, faces, anywhere they may have
brushed against the little pests. Neighbors arguing angrily about
whether to resort to risky chemicals. Noisy town meetings. Anguished
editorials in newspapers.*

Science magazine quoted a federal entomologist to the effect
that the gypsy moth has "won the war." Reviewing past outbreaks, the reporter for *Science* noted that every time the moths
declined, so did appropriations, setting the stage for a new outbreak. Following the devastation of 1981, increased funding
became available, but "there is really no expectation of stopping
the moth's progress."

In California, the number of moths trapped in the San Francisco area increased from one in 1975 to five in 1980, 59 in 1981,
and 102 in 1982. This has led the state Department of Food and
Agriculture to initiate aerial spraying of limited areas with Sevin
despite protests by the Sierra Club and other environmental
groups claiming that other alternatives had not been seriously
considered.

The despair of citizens seems not to be reflected in the Washington bureaucracy. A Gypsy Moth Planning Task Force
(GMPTF) has recently been appointed, to include federal and
state officials and representatives from industry and the academic community. The objective is to prepare a Comprehensive

Gypsy Moth Program Proposal (CGMPP). We live in an age of acronyms! Too bad these are unpronounceable; surely GYP-MOP would be more inspiring ("Mop up the gypsies!"). But I wish them success. Cognizant of recent advances in novel control methods, and using computer modelling, optimization theory, cost/benefit analysis, and something called the "purposes-target-results" approach, they may succeed without ever having to look at a gypsy moth.

Why is the gypsy moth proving so tough to subdue? It is a complicated beast, and not all individuals are by any means alike. Depending upon what and how much they have eaten, what temperatures and moisture they have been exposed to, and how crowded they have been, individuals vary in rate of development, ability to disperse, susceptibility to disease, and reproductive capacity once they have reached adulthood. Crowding may cause individuals to molt more times than normal and may even cause slight color changes. When populations are increasing, there tends to be a preponderance of females, and these lay larger than normal egg clusters.

At a population peak, females tend to lay smaller eggs with less yolk, and these produce small larvae that have a greater tendency to disperse. At high population densities, larvae feed during the day, abandoning their usual behavior of feeding at night and resting in hiding during the day. The trees that are being fed upon also assert an influence on their defoliators; the leaves undergo nutritional changes that influence the growth rate and vigor of the caterpillars. Thus there is a subtle interplay between insect, plant, and the elements.

It is also clear that not all populations of gypsy moths are alike. This is well known in Europe and Asia as a result of the studies of Richard Goldschmidt, who in 1934 produced a 185-page monograph on the subject. He found that certain populations differ from others in color of the larvae, rates of development, number of molts, egg size, duration of winter rest, and even size of the chromosomes. Crosses between individuals from different areas often produce sterile individuals. In Russia, there is even a race of gypsy moths in which the females disperse by flight, in contrast to the sluggish, nonflying females in other parts of the range.

Even though North American gypsy moths come from at most two introductions from Europe, they are by no means all alike. David Leonard, of the University of Maine, has recognized several strains that differ in rate of development and resistance to cold. In most areas, females lay their batches of eggs at various heights on tree trunks, but at the northern edge of the range, most are laid within five feet of the ground, where they stand a good chance of being shielded from the cold by snow. Most hatching, in fact, occurs below one foot, where the snow cover is deepest and most dependable. In Russia, there are populations that consistently lay their eggs close to the ground, where they survive air temperatures more than 50 degrees below zero Fahrenheit by taking advantage of the snow cover. It is not surprising that so adaptable an insect is able to subvert our efforts to control it. Michael Gerardi and James Grimm conclude their book on the gypsy moth with the comment that "it may be that the biology and population dynamics [of the insect] are too well evolved for present technology to control."

Perhaps we had better heed the advice of naturalist Roger Swain, who wrote in the *New York Times Magazine*, "we'd best learn to live with them." As he pointed out, foresters tend not to worry a great deal about gypsy moths, as over time they at most cause the death of already weakened trees and to some extent cause trees less favored by the moths (such as beeches and maples) to replace more favored trees (chiefly oaks). Red maples even profit by the increased light and nutrients derived from caterpillar droppings. Long-range changes in forest composition are always going on, and gypsy moths produce nothing like the changes caused by human intervention.

As Roger Swain puts it, the gypsy moth is "not really a pest of trees, it's a pest of people." No one enjoys having his shade trees stripped of their leaves, or of having hairy caterpillars all over his yard, at times causing skin rashes and a repulsion to the entire world of insects. But it is in suburban areas that the broad-scale application of chemical poisons arouses so much opposition, and with good reason. Perhaps the answer is not high-sounding acronyms, but a packet of instructions, parasites, disease spores, and pheromone dispensers that can be made available to individual householders. Then everyone can be-

come an instant entomologist and enjoy the excitement and
frustration of coming to terms with some of the earth's most
successful organisms.

7

"Killer" Bees

Insects have always had a bad press. Outbreaks of pests regularly make our newspapers and television newscasts, but the virtues and attractions of insects are rarely discussed. By and large, the nuisances and dangers posed by insects are grossly exaggerated. A yellowjacket on the patio, aphids on the nasturtiums, an unidentified speck in the soup: all are cause for alarm.

Most people just don't like "bugs," I am afraid, even though as children they may have marveled at a beetle or a butterfly. Sooner or later, most children are taught by their parents, their teachers, or by the communication media that insects are not "nice": they bite, sting, itch, eat our crops, carry diseases, and so forth. They are never told that fewer than one percent of insect species are "pests" in any sense and that most live lives that seem obscure to us but are fundamental to the health of the environment. As to what is a "pest," it is very much a matter of one's training and point of view. Dislike of insects reaches its ultimate in cases of entomophobia, a nervous condition in which one imagines oneself constantly harassed or bitten by insects.

The media had a field day when honey bees of an aggressive African race became established in Brazil and began to threaten us from the south. "Killer bees," they were called, and stories were told of their deadly attacks on people and livestock. For a time a magazine was on the newsstands called *Killer Bees*. It was filled with scare stories as well as sketches of people fleeing from bees or threatened by ants, scorpions, spiders, and other "life-threatening" creatures. Even the staid *National Geographic* ran an article entitled "Those Fiery Brazilian Bees," providing an essentially accurate account sprinkled with tales of bees attacking people, horses, and cattle. For example, a Brazilian teacher, Dr. Eglantina Portugal, was reported to have killed a bee in her school, causing it to release alarm pheromone and attract bees from nearby swarms. She ran, but fell down and was overcome by bees. In the words of a witness: "Some people from nearby houses arrived with water, but the bees stung them and drove them back. Finally, firemen arrived, but they, too, ran away with bees all over them. They returned waving smoking torches and were able to get Dr. Eglantina to the hospital. But it was too late."

I have no reason to doubt this story or many of the others that have been told of the ferocity of Brazilian bees. Nor do I want to imply that we can always afford to take a casual attitude toward insects: at times they do get out of hand and warrant our attention. Orley Taylor, of the University of Kansas, has spent a number of years, some of them in South America, studying these bees. According to him, the bees "are *not a major* public health hazard [his emphasis]. The number of people affected in stinging incidents is likely to be small compared with numbers affected by other health problems." While the problem cannot be shrugged aside, it pales by comparison with mortality to humans caused by our worst enemy: ourselves. However fierce, no insects can compete with terrorists, warmongers, or for that matter speeding or drunken drivers, when it comes to directly destroying human life (insects' occasional role as disease vectors is another matter). We live in a society in which we reluctantly condone many kinds of death and debilitation when self-inflicted. But to be stung by a bee!

A matter of more concern to North Americans comes from quite another attribute of Brazilian bees. Studies of their ecology and behavior suggest that there may be nothing to prevent them from eventually reaching the United States. However, they may

The honey bee is the most familiar of all insects, though still far from fully understood.

P. EADES

never persist beyond the southern tier of states because they do not form winter clusters inside the hive, like the European honey bees in common use, so they cannot survive prolonged cold. But most American beekeepers start or requeen their colonies from stocks purchased from bee breeders in Texas and other southern states. If wild colonies of Brazilian bees become established there, they may contaminate stocks of European bees by interbreeding with them. Thus queens supplied by these breeders may have attributes of the Brazilian race: they may be more aggressive, they may swarm more readily, and, most important, they may not survive the winters.

Some 200,000 people in the United States keep honey bees. Professional beekeepers not only market honey and beeswax, valued at $140 million annually, but the use of colonies for pollination of orchards and field crops is another major source of income. The annual value of United States crops pollinated by insects (chiefly honey bees) exceeds 10 billion dollars. Beekeeping is also one of the finest of hobbies, but hobbyists are not likely to put up with unfriendly bees, especially if they have to restock their hives every spring.

Note that I have been calling them "Brazilian" bees, following recent practice. The original stocks came from Africa and belonged to a race well adapted to the warm temperatures of that continent. After they escaped in Brazil, they began to interbreed with the European bees in use there, producing a hybrid race that still retained many of the features of the African bees but showed much of the variation often characteristic of uncontrolled hybridization. The terms "Brazilian" or "Africanized" are meant to imply that the bees are no longer of pure African stock. What the ultimate result of this interbreeding will be is uncertain. One may hope for a breed that is more mild-tempered but, like the African bees, a superior honey producer.

This was the original hope of Dr. Warwick Kerr, of the University of São Paulo, who in 1956, at the request of the Brazilian government, imported African queens (no connection with C.S. Forester's novel *The African Queen*, or with the movie based on it!). European bees had never proved outstanding honey producers in the tropics, so it was hoped that a careful breeding program might produce a stock with the best attributes of both races. The breeding was to be done in an isolated area, using artificial insemination of queens by the semen of carefully selected European drones. Until a safe and satisfactory stock could be bred, the hives would remain shielded by grids that permitted the workers to pass in and out, but had slots that were too narrow to permit the free passage of queens and drones. In a scenario paralleling the introduction of the gypsy moth to Massachusetts nearly a century earlier, some of the queens did in fact escape with swarms and establish wild colonies. Evidently a visiting beekeeper did not understand the potential problems and removed the grids.

Within a short time, the African bees had begun to interbreed with and replace the European bees kept by local beekeepers. There was great alarm at the ferocity of the bees, and many beekeepers went out of business. By 1970 the bees had spread over most of Brazil, and by 1978 they had reached Venezuela and Colombia, apparently poised to invade the Central American countries one by one. The "domino theory" so often applied to the supposed spread of communism seemed now appropriate to "killer" bees!

Responding to the public outcry, Warwick Kerr began supplying beekeepers with queen European bees in the hope of diluting the aggressiveness of the Brazilian bees. More than 23,000 queens were distributed, and they did seem to result in more docile strains in southern Brazil. But in other areas, the bees were plentiful in the wild, nesting not only in hollow trees but in unusual places such as holes in the ground and in termite nests. In one area Kerr counted as many as 107 colonies of wild bees in one square kilometer, a remarkably high number. Furthermore, Brazilian bees are forever on the move, forming swarms much more frequently than European bees and sometimes migrating considerable distances. In northeastern Brazil, where European bees had never been abundant, the Africanized bees seemed especially aggressive, and it was these bees that seemed to be spreading along the coast toward Central America.

From the beginning, there was much disagreement over the dangers posed by the bees, over the difficulties in managing them, and over their superiority as honey producers. Some of this disagreement was doubtless caused by real differences in the genetics of local populations, but factors such as weather and abundance of nectar sources probably played a role in influencing their "disposition." How "fierce" the bees are is often a subjective matter; even the tamest of bees frighten some people. In order to try to quantify aggressiveness, researchers have invented a "sting recorder," consisting of a ball of black leather which is jiggled for sixty seconds at various distances from a hive. Jiggled at a distance of twelve inches from the hive, such a device attracts an average of two stings from European bees, but an average of fourteen stings from Brazilian bees. The latter will also pursue a leather ball a much greater distance—on the average about 500 feet for Brazilian bees as compared to 70 feet for European bees. One researcher was pursued over half a mile, the bees trying to sting the leather ball all the way. When bees sting, they release a pheromone that causes other bees to attack. Brazilian bees appear to be much more responsible to this pheromone than European bees.

In appearance, Brazilian bees differ little from European bees. They are slightly smaller and darker, and the combs they build have smaller cells. Most of the differences relate to their behavior. They respond quickly and *en masse* to disturbances to their hive; they reproduce rapidly and send out swarms every few months; they abandon their hives more readily than European bees if conditions are not just right. They also fly faster than European bees, rendering them less able to be caught by certain birds and wasps that specialize on bees. In most areas they are superior competitors for nectar as compared to European bees and to various native bees. They even differ in details of the "waggle dance," used to communicate distance and direction of food sources to nestmates. Karl von Frisch had said that the waggle dance was used if the source was more than about eighty meters away; but Brazilian bees use the waggle dance if the source is as little as twenty meters distant.

The many diverse reports issuing from South America—many of them quite lurid—caused considerable alarm among North American beekeepers and in fact among the general public. In Orley Taylor's words, "The many sensationalized stories in the popular press have made it difficult for anyone to obtain a proper perspective of past and possible future events." In the effort to clarify the problems, the National Research Council established an investigatory committee of nine members, drawn from bee experts in the United States, Canada, and Mexico. The committee visited Brazil and in 1972 submitted a report on their findings. The committee chairman, Charles D. Michener, of the University of Kansas, summarized some of their findings as follows:

> *If the Brazilian bee reaches the United States in a form similar to that now found in Brazil, it will have a substantial impact. There will be the public health matter—stingings of people. There will be similar problems with domestic animals—dogs, horses, chickens, etc. At least initially, it will have an adverse effect on honey production. The impact could also include a decrease in our use of the out-of-doors and of people's pleasure there. Because the queen rearing and packaged bee industries are in the southern states, its impact will be continent-wide, not limited to the warmer states where the Brazilian bee is likely to be able to replace European bees.*
>
> *Economically the most important influence of the Brazilian bee is likely to be on our pollination industry which has almost no counterpart in Brazil. So far as is known the Brazilian bee is a perfectly good pollinator, but it will not be practical to place hives in fields and orchards where people and machinery are working.*

The report went on to recommend several ways in which the arrival of the Brazilian bees might be delayed or prevented. Bees in a barrier zone in Central America could be systematically destroyed so as to prevent the passage of "African" genes northward. Since the barrier zone would undoubtedly contain many wild colonies in forested areas, it might be possible to drop biodegradable trap boxes from the air. The idea was that many of these would be occupied by the bees but would have sufficient insecticide in their walls that chronic exposure would kill the bees. Or, the spores of certain known diseases of honey bees might be disseminated in the barrier zone. Still another method might be the distribution of poison bait from planes. The bait would consist of a sugar solution containing tiny capsules resembling pollen grains which are, in fact, capsules of poison.

These measures have an air of desperation about them, and understandably none of them have been carried out.

Several South American bee experts, including Warwick Kerr, took issue with some of the conclusions of the committee. They pointed out that bees in southern Brazil had become less aggressive and that beekeepers had learned to modify their techniques so as to take advantage of their superior performance in honey production. In some areas, the harvest of honey and beeswax had greatly increased. True, bees had become unpopular with hobbyists, and not everyone agreed that there was a noticeable decline in aggressiveness. But those who kept bees for profit were mostly happy with their Brazilian colonies. They had learned to scatter their hives, away from houses and barns, and to use large smokers to subdue the bees when they needed to work with the hives. In Bolivia, even laughing gas had been used to anesthetize the bees while they were being worked. Beekeeping has a long tradition, extending far into antiquity. Beekeepers have been slow to learn new ways, and this would surely be true if more aggressive strains were to reach North America. But whether the Brazilian bees will reach the United States and how we will respond to them remain open questions.

In any case the popular press seems to have abandoned "killer bees." Evidently there have been enough murders, terrorist attacks, and threats of nuclear holocaust to fill its pages. One wonders what that poet of the bees, Maurice Maeterlinck, would have to say of "killer" bees. Familiar only with relatively docile European honey bees, he described the sting as "a kind of destroying dryness, a flame of the desert rushing over the wounded limb, as though these daughters of the sun had distilled a dazzling poison from their father's angry rays, in order more effectively to defend the treasure they gather from his beneficent hours."

But, he went on, "the slight amount of skill needed to handle [the bees] with impunity can be most readily acquired." Not so with Africanized bees. Perhaps early races of mankind, as they evolved on the African plains, made use of honey as a major source of carbohydrates. Over several million years, natural selection may have produced a race of bees unusually able to defend themselves and to move readily from one place to another. Perhaps their superior qualities as honey producers in the tropics might have been bred into more docile stocks had not an unfortunate accident occurred. But as it stands, we shall have to wait and see what happens.

Although honey bees are sometimes said to be our only "domesticated insects," even the gentlest of bees are far from exhibiting the vacuities of our livestock or the subservience of our household pets. It is sobering to consider that such fragile, small-brained creatures as insects cannot readily be bent to our wills.

8

Blister Beetles

There are some insects that even an entomologist finds it hard to love. Among these are the blister beetles. Most are rather drab in color and soft in texture, not hard as any self-respecting beetle ought to be. These beetles live up to their name by producing large and annoying blisters when handled. They also produce a rather fetid, mousy odor. The beetles usually occur on plants in aggregations, and their aggregative behavior results in the production of an odor that can be detected some distance away, doubtless an effective deterrent to predators.

F. E. L. Beal, an entomologist formerly with the U.S. Department of Agriculture who has the unenviable reputation of perhaps having studied the contents of more bird stomachs than anyone else, found only one blister beetle in 6,184 stomachs of crows, starlings, and vireos. In 1930 and 1931, naturalist Frank Morton Jones put out hundreds of freshly killed insects in feeding trays outside his home on Martha's Vineyard, Massachusetts. Ninety percent of the beetles were consumed by various birds, but fewer than 10 percent of the blister beetles, and these mostly by young, inexperienced blue jays. Sunfish, too, are known to spit out blister beetles, although they consume most insects readily. So there seems little doubt that the noxious properties of these beetles serve them well against vertebrate predators.

Thomas Eisner and his colleagues at Cornell University have shown that blister beetles are equally well protected from the attacks of other insects. When an ant attacks and grasps a leg, the beetle discharges a droplet of blood from a joint of the leg, and a continued attack by several ants results in bleeding from several leg joints. Ants contaminated with blood quickly run off and clean themselves by rubbing their bodies against the soil. Ground beetles, which are voracious predators on most insects, refuse to attack blister beetles, but when the beetles are literally stuffed into their mouths they, too, run off and rub their mouths in the dirt. Eisner showed that when extracted blood of blister beetles is added to sugar water, even in minute quantities, ants do not readily accept it.

The substance that causes avoidance by predators is evidently the same substance that causes blisters in humans. It is called cantharidin, and has been known for a very long time. Hippocrates, in the fourth century B.C., is said to have recommended extract of blister beetles as a cure for dropsy. According to Lucy W. Clausen, in her book *Insect Fact and Folklore,* Frederick the Great offered a sum of money for the development of a cure for hydrophobia. The cure that was developed called for capturing twenty-five blister beetles, slipping a hair around their necks, and hanging them up until dry. They were then preserved in honey to which snakeroot, lead filings, and other substances were added. Doubtless this concoction was administered with suitable incantations.

Kirby and Spence, in 1815, referred to blister beetles as "the only insects which directly supply us with medicine." A European species of beetle sold at sixteen dollars a pound was considered "of incalculable importance as a vesicatory." The production of blisters was considered to provide relief from internal disorders, somewhat in the manner of a mustard plaster. In more recent years, cantharidin has been used as an ingredient in hair restorers. As late as 1940, it was still listed in the pharmacopoeias of England, Belgium, Spain, and Italy.

But it is as an aphrodisiac that cantharidin is best known, generally under the name "Spanish fly." Hippocrates recognized that it was an irritant to the sex organs, and the Romans are said to have used it to prolong their orgies. The dried, powdered beetles (or their blood) are evidently effective, and are said to produce painful and prolonged erections in human males.

Listening to the ethereal music of Wagner's *Tristan und Isolde*, it is disheartening to realize that the plot hinges on a dose of Spanish fly slipped by Brangäne to the ill-starred lovers.

In parts of Europe, in days gone by, blister beetles were harvested for the production of Spanish fly. Heavily clothed, with gloves and face masks, the harvesters set out early in the morning, while the beetles were still sluggish with cold, and shook them into cloths placed on the ground beneath the bushes. Then, to avoid losing any of the secretion, the beetles were plunged into vinegar or placed in sieves and exposed to the fumes of vinegar, ammonia, or sulphur. They were then dried in the sun or in ovens and eventually pulverized to form the major ingredient of love potions. Animal breeders also have used Spanish fly to excite reluctant boars and stallions.

Even in very small quantities, cantharidin is a powerful poison. As small a dose as half a milligram per kilogram of body weight is lethal to humans, and even traces can damage the kidneys severely. In recent years, there have been reports of deaths to cattle and horses that were fed hay or silage containing blister beetles. The use of cantharidin in medicine or as a sexual stimulant has fortunately ceased in most parts of the world. However, there is no evidence that the demise of Spanish fly has slowed the world's population growth, or its fascination with sex.

There are a great many kinds of blister beetles. Certain species show a preference for feeding on certain kinds of plants; one is called the "old-fashioned potato beetle," evidently because it was a pest of potatoes before the infamous Colorado potato beetle usurped its major role. As a group, blister beetles attack a wide variety of wild and cultivated plants and at times do a good deal of damage by eating the leaves and flowers.

The larvae of blister beetles lead much more obscure lives than the adults, and they were little known to the ancients. As late as 1841, Thaddeus W. Harris wrote, in his classic *Treatise on Some of the Insects Injurious to Vegetation* (which I shall say more about in a later chapter), "the larvae are slender, somewhat flattened grubs [which] appear to live upon fine roots in the soil." This statement was repeated in the 1862, posthumous edition of Harris's book, although by that time it was clear that blister beetle larvae do not live on plant tissue at all.

The larvae of some species are destructive in quite a different, unexpected way. They destroy the nest cells of various ground-nesting wild bees which play a role in the pollination of wild flowers and cultivated plants. The adult beetles, however, have no obvious association with bees. How their larvae get to the bees' nest cells is a story well told by Fabre (though most details of the story were known before Fabre's time). Fabre's essay was titled "The Oil-Beetle's Journey," oil-beetle being a common name for certain types of blister beetles that produce an abundance of an oily fluid when disturbed.

The minute, recently hatched larvae of blister beetles were in fact known to Linnaeus and to Réaumur; the latter found them attached to the hairs of bees and gave them the name *Pediculus apis*, meaning "bee louse." It was Leon Dufour who dubbed them "triungulins," meaning "three-clawed," since the single claws are flanked by two bristles (in some species), giving the appearance of three claws. Early workers can be forgiven for calling them lice, for they are indeed remarkably louselike, and it remained for the British entomologist George Newport, in 1851, to identify them as new-born blister beetles, in a classic series of papers published by the Linnaean Society. How do the larvae get to the bees, and what happens when they are carried to the bee's nest? This was the gist of the "journey" described by Fabre.

Fabre had been lying in the grass watching the activities of a group of Anthophora bees—bees that build mud turrets outside their nests, which are dug into vertical banks. He found himself covered with "little yellow lice," which he recognized as blister beetle larvae. Others were seen on the ground "running in a great flutter" and climbing up and down grass blades. Still others had collected on the flowers of plants growing nearby. Remembering Newport's discoveries, he concluded that adult oil-beetles had laid their eggs on the ground near the bee nests and that the newly hatched larvae were finding their way to the tops of flowering plants, there to attach themselves to bees and be

Blister beetles' lives are much more complex than their drab appearance would suggest.

P. EADES

carried to the nest.

Fabre caught several bees and found that nearly all of them bore larvae. So did many other flower-visiting insects, such as wasps and flies. He put a spider within reach of the larvae, and they readily climbed on, as they did also with a piece of cloth. What eventually became of larvae that attached themselves to an object other than an Anthophora bee? Of course they perished. "Instinct is at fault here," Fabre wrote, "and fecundity makes up for it." Blister beetle females are known to lay great numbers of eggs—several thousand in some cases—leaving the small, active larvae to shift for themselves.

"But instinct recovers its infallibility in another case," Fabre continued. On the flowers, the larvae remain motionless and somewhat hidden until there is a disturbance, whereupon they scatter to the tips of the petals, where they hang with their legs free, ready to grasp a visiting insect. On such an insect, they remain motionless until carried to a nest. But on a piece of cloth or tuft of cotton they move about actively, evidently sensing that they have "made a mistake" and must regain the flowers. How do they distinguish between the hairs of an insect and those of a piece of cloth?

"Is it by touch," Fabre asked, "by some sensation due to the inner vibrations of living flesh? Not so, for the triungulins remain motionless on insect corpses that have dried up completely, on dead Anthophorae taken from cells at least a year old. . . . By what sense then can they distinguish the thorax of an Anthophora from a velvety pellet, when sight and touch are out of the question? [The larvae are essentially blind.] The sense of smell remains. But in that case what exquisite subtlety must we not take for granted? Moreover, what similarity of smell can we admit between all the insects which, dead or alive, whole or in pieces, fresh or dried, suit [the triungulins], while anything else does not suit them? A wretched louse, a living speck, leaves us mightily perplexed as to the sensibility which directs it. Here is yet one more riddle added to all the others."

Once having attained the nest-cell of an Anthophora, by this precarious route, how does the larva avoid becoming drowned in the nectar in the cell and how does it avoid having to compete with the bee larva for food? This was a riddle Fabre did solve. When the bee lays her egg on top of a mass of pollen and nectar, the larva drops from the bee onto the egg "which serves it for a raft." The larva first consumes the bee's egg, growing and assuming a different body form, able to penetrate and consume the mass of food in the cell.

Many of the observations of Newport and of Fabre were confirmed in 1924 for an American species of blister beetle that attacks quite a different ground-nesting bee. J. B. Parker was a professor at Catholic University in Washington, D.C., and made his observations on the campus of that university—not the first instance of field research emanating from the heart of a major city. Parker clarified a number of further points. For example, what happens if several triungulins drop off onto a single bee egg, as must often happen? Simple: they devour one another, until only one remains. And what happens to those that attach themselves to male bees, which do not build or provision cells? Since male bees emerge before the females and spend much time on flowers or on the ground looking for females, they do indeed pick up many triungulins. Parker found them mainly on the under side of males, but mostly on the upper surface of the females. He surmised that when the male mounts the female during copulation, the triungulins pass from male to female.

Parker's article was published jointly with Adam Böving, who provided detailed descriptions of the various stages of the larvae, for in fact the larvae, as they develop, transform into several quite different, successive forms. These larvae are often cited as prime examples of "hypermetamorphosis," which means simply that they are "hyper" with respect to change in body form. The active, long-legged triungulins transform into robust, nonfeeding "pseudopupae," which transform into still another body form before molting to the true pupal stage.

Böving, a skilled artist, rendered each body form with as much loving detail as another artist might devote to the human body. He was Danish, educated at the University of Copenhagen, but he moved to the United States in 1913 and worked for the U.S. Department of Agriculture until he retired in 1939. He and a colleague, F. C. Craighead, published in 1931 an *Illustrated Synopsis* of the larval forms of all beetles, a monumental work that is still a standard reference. Publication of the 125 plates of illustrations was an expensive proposition, and Böving is said to have paid for most of the costs himself.

So it turns out that blister beetles are rather more complicated animals than one might suppose. But I have not told the whole story. While many do in fact live in the cells of bees in their immature stages, as Newport and Fabre had described, the triungulins of some species show no inclination to cling to the hairs of bees but walk over the ground and dig in here and there. Discovery of what they were up to is credited to Charles Valentine Riley, a brilliant and colorful American entomologist whose involvement with the infamous Rocky Mountain locust is a story I told in my book *Life on a Little-Known Planet*. While digging up egg masses of the locust, in the 1870s, Riley found pseudopupae in some and in others young larvae devouring the eggs. Eventually, he reared three species of blister beetles from these larvae. In some places a high percentage of the egg masses of this and other species of grasshoppers had been destroyed by blister beetles. So some of these rather repulsive and destructive insects suddenly emerged as friends of man.

Riley was able to observe females digging little holes in the ground, laying masses of eggs in them, and scratching soil over the holes with their feet. Since they prefer to lay their eggs in much the same places as grasshoppers, the triungulins do not have far to travel, though doubtless many perish without finding food. Riley wrote of them: "At night or during cold weather all of those in a batch huddle together with little motion, but when warmed by the sun they become very active, running with their long legs over the ground and prying with their large heads and strong jaws into every crease and crevice in the soil, into which, in due time, they burrow."

Once having found a grasshopper egg pod, the triungulin burrows in and feeds on an egg, then proceeds to the next egg and, after a series of molts and transformations not unlike those of species that attack bee cells, may eventually consume most of the eggs in the pod, sometimes fifty or more. The adult beetles that are finally produced then feed innocently on foliage, providing no evidence of their very different life beneath the soil. Like other blister beetles, they are quite capable of defending themselves chemically. Many of the species that attack grasshopper eggs belong to a genus appropriately named *Epicauta,* literally "a burn on the outside" (as in the word cauterize).

Adult blister beetles, too, have diverse structural peculiarities, particularly among the males. How these might be related to differences in mating behavior is a problem attacked several years ago by Richard Selander of the University of Illinois. He and his students have devoted a great deal of effort—and doubtless had a good deal of fun—exposing the intricate details of the sex life of blister beetles. Their findings are not likely to influence the GNP, but they are rewarding enough in terms of the PGN (profounder grasp of nature), which in the long run is going to have a lot more to do with human survival than the GNP.

A female blister beetle is not a very prepossessing creature from a human point of view, but when males are crowded with females in a cage, they undergo what Selander calls a "mating frenzy." Competition for females is so severe that males cannot complete a courtship and copulation without interference from other males, and they become so stimulated that they attempt to mate with other males, or even with the walls of the cage. In nature, in one group of species studied by Selander, the male approaches the female from behind and mounts her fully, so that his head is close above hers. Then he vibrates his antennae against her body and begins to touch her repeatedly with his curiously modified palpi (short, antennalike structures associated with the mouthparts). "In anthropomorphic terms," Selander says, "it seems that the male is intent on antennating and palpating, in a minimum of time, as much of the body of the female as is accessible to him. . . . " At the same time his body rocks forward and backward. After a short time he slides backward and assumes a position behind the female, still using his antennae and palpi and also grasping her hind legs with his front legs. Then he strokes her abdomen with specialized brushes on his abdomen and begins to insert his genitalia.

For the first time, the female is usually not sufficiently docile to permit copulation, so the male remounts and repeats the cycle, sometimes several times before she is submissive. When he finally succeeds in inserting his genitalia fully, he falls over backward and then rights himself, so that he is facing in the opposite direction of his mate. This causes him to rotate his body 180

degrees with respect to his genitalia. Then, for about ten minutes, he pulsates his abdomen, passing sperm to the female. The
pair remain locked together for several hours, during which
both may feed, and the female may be courted by other males.

While all the species in that particular genus *(Pyrota)* do
much the same thing, there are differences in the sequence of
various acts among the species that have been studied. In another genus, *Epicauta,* the males of many species have the basal
segment of their antennae greatly enlarged. The significance of
this was understood when Richard Selander and his students
discovered that males of these species engage in what they call
"antennal whipping" and "antennal wrapping." Again there
were differences in detail among the various species studied,
differences doubtless associated with preventing copulation
with alien species in nature. This is only one of many examples
of the way that studies of behavior may help to explain structure—why animals are built the way they are.

So blister beetles, those drab, rather stinky denizens of our
gardens, in fact are rather remarkable bundles of adaptations,
defended by poisonous substances in their blood, engaging in
elaborate courtships, and producing great numbers of eggs that
give rise to delicate louselike larvae that face the hazards of finding their way to a very special source of food, either grasshopper
eggs or the cells of certain bees. Enough to stimulate our admiration, but perhaps not our love. And enough to demonstrate how
hard it often is to label insects as friend or enemy of the human
species. Personally, I wouldn't want to part with any of them, so
long as we understand them well enough to live in peaceful
coexistence.

9

The Medfly

One of the joys of our modern technological world is our ability to travel quickly and comfortably to almost any part of the world. A boon to insect aficionados surely, for they can (if they are able to afford it) spend a vacation in Brazil chasing morpho butterflies, or in Africa hunting goliath beetles. And a boon to professional entomologists, who meet in international congresses every few years to discuss common problems. In fact, as a result of modern commerce, most insect problems are international in scope. Pests such as the gypsy moth, codling moth, pink bollworm, and many others, have been disseminated to many parts of the world. There are those who claim that in another century much of the earth's fauna and flora will be international, restricted only by latitude and by rainfall patterns. That is, most of the plants and animals of tropical forests that still exist will occur throughout the tropics, most temperate crop plants and their pests will occur everywhere that conditions permit, and so forth.

Already, in tropical areas, it is hard to escape lantana, coconut palms, and the like; eucalyptus trees have been planted extensively on all continents, though in their native Australia replaced over broad areas with pines from the northern hemisphere; and the chant of house sparrows can be heard in cities throughout the world. Insects, with their powers of dispersal and their ability to conceal themselves in shipments of foodstuffs, nursery stock, or almost anything, are rapidly becoming "citizens of the world" in a more real sense than the human species seems capable of.

Consider the Mediterranean fruit fly, which occurs pretty much throughout the tropics and subtropics of the world. As of this writing, it probably does not occur in the continental United States, but it has made its appearance here at least six times, each time producing headlines in newspapers and furious activity on the part of entomologists. (The name Mediterranean fruit fly is much too long for newspaper headlines, hence the popular name "medfly.")

Evidence suggests that the native home of the medfly is in tropical West Africa. It was first noted in the Mediterranean area about 1820, and in 1829 it showed up in oranges that had been shipped to London. In 1897 it was reported from Australia, presumably having arrived in imported fruit. By 1900 it had invaded South America, where it soon devastated orchards of peaches and other fruits.

In 1910, on June 21, to be precise, entomologist D.T. Fullaway captured a medfly in the U.S. Department of Agriculture insectary in Honolulu. Soon it was discovered in citrus orchards on Oahu, and within a few years it was found to be present on all of the major islands of Hawaii, attacking oranges, limes, peaches, guavas, mangoes, coffee berries: in fact, no less than seventy-two kinds of fruits. Since there is no time in Hawaii when fruits of some kind are out of season, the medfly found no shortage of food and produced successive generations throughout the year, sometimes as many as fifteen. It flourished not only in orchards but in cities and towns, where there were plenty of fruit trees and fruiting ornamentals around homes, and even in the wild, where there were plenty of wild guavas.

The medfly is particularly dangerous as it is a "direct pest," that is, it attacks that portion of the plant that is consumed by humans, namely the fruits (in contrast, for example, to potato beetles, which attack the plants but not the part we eat, the tubers). Even a minor blemish on a fruit renders it unsalable, so conditioned are we to perfect fruits. But the medfly does not stop at minor blemishes. The maggots feed throughout the fruit, often reducing it to a spongy mass and permitting the development of fungi and bacteria. Sometimes it is difficult to judge from the outside how much damage has been done, since the effect of egg-laying punctures varies depending upon the kind of fruit and its maturity. Often hardly anything is visible from the outside. Oranges squeezed for juice may contain small maggots that pass as pulp. The maggots can not survive the digestive juices of adults, but there is evidence they can cause temporary damage in the digestive systems of small children. When medfly is known to be present, all fruit is under suspicion!

The medfly is a truly beautiful little animal, about the size of a house fly but with reddish eyes, a body streaked with yellow, black, and silver, and wings spotted and banded with yellow, brown, and black. What could be the function of such psychedelic patterns, far more elaborate than is surely necessary for the flies to recognize other flies as members of their own species? These patterns are, in fact, part of an elaborate sexual display. Over the years, those individuals with the most effective displays have been most successful in mating and producing offspring. Thus, evolution has resulted in an intensification of these displays (as it has in many insects). In the medfly, colors are only part of a display that includes movements, sounds, and subtle odors.

Much of what we know of medfly courtship and mating is a result of studies by Ronald Prokopy, of the University of Massachusetts, and Jorge Hendrichs, of the Direccion General de Sanidad Vegetal, Mexico City. These researchers placed cages over wild medflies on coffee trees, where they could monitor mating success under various conditions. During late morning and early afternoon, males gather on the underside of leaves in small groups, called leks, where they pool their resources—especially their sex pheromones—to attract females. Each male maintains a small territory on the underside of a leaf, about a foot away from another male, and when a female arrives she selects a mate, perhaps on the basis of the vigor of his posturing, the quality of his pheromone, his wing sounds, or the visual impression made by the sum of his color spots and bands; or, more likely, by the combination of all these stimuli.

When the female has selected a mate, she walks toward him until she is only a few centimeters away. The male then fans his wings in a particular rhythm, producing a high-pitched hum and directing his pheromone toward her. When she is sufficiently mesmerized, he jumps on her back and attempts to mate with her. Often he fails the first time; one pair was seen to go through the courtship routine five times in thirty minutes before copulation occurred. Before the male is able to pass sperm to the female, he must clasp and draw out the tip of the female's abdomen, at the same time uncoiling his long, ribbonlike copulatory organ, which loops about until the tip reaches her genital orifice.

It is usually virgin females that are attracted to the leks near midday, and the majority of these mate there. At other times of the day, the males move to the fruits, where they encounter nonvirgin females laying their eggs. Here they largely dispense with their displays and attempt to "rape" the females. Although they succeed only once in a while, when they do succeed their sperm may replace that of another male (since in most insects it is the last sperm received that fertilize the eggs). So it is to the male's advantage to have these two alternative mating systems; in this way he may father more offspring.

The flies sometimes live for several months, mating frequently. When no suitable fruits are available, females are able to wait for long periods and begin laying eggs whenever they can find a place to lay them. Eventually, a female may lay several hundred eggs. Each time she lays a small batch of eggs, she afterward drags her egg-laying tube over the outside of the fruit, marking it with a pheromone that persists for a week or more. In this way other flies are informed that the fruit has already been attacked, and they would do better to lay their eggs elsewhere.

The eggs hatch in a few days and the maggots bore into the fruit and begin to feed. They grow rapidly, reaching full size in ten to forty days, depending on the temperature, then leaving the fruit and moving to the ground. Here they move about for a time, looking for a crevice or soft spot where they can enter. They are remarkably adept jumpers, fastening their mouthhooks onto the back of their body, like a cat biting its tail, then letting go suddenly with a snap that sends them several inches into the air. According to one report, a maggot can cover up to fifteen feet in ten minutes—rather a neat trick for an insect

---

The medfly—a thing of beauty but hardly a joy to orchardists.

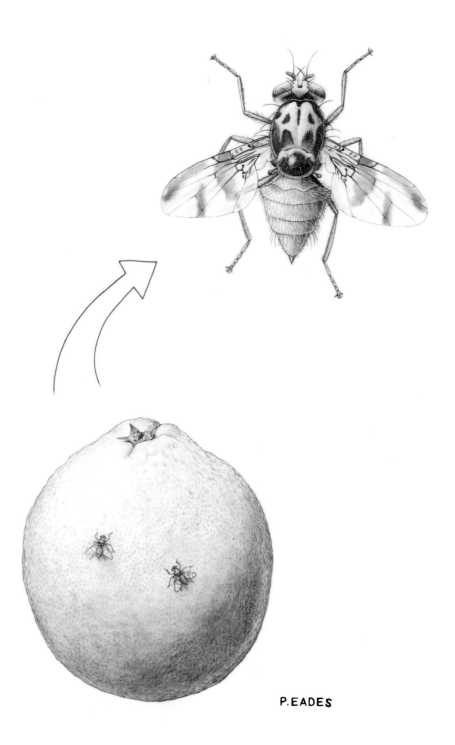

P. EADES

that has no appendages at all.

When the medfly arrived in Hawaii in 1910 it carried with it a considerable reputation as a destroyer of fruits of many kinds, and there was fear that it would not only become a serious pest in Hawaii (which it did) but that it would have a jumping-off place for the continent (which it also did). The Hawaiian Board of Agriculture and the U.S. Department of Agriculture set about to see what could be done about containing the pest. Infested fruits were gathered and destroyed by burial or incineration, as were wild or nonessential fruit trees. Fruits were even dropped into the sea in weighted sacks. But inspectors ran into problems convincing homeowners that their fruit had to be destroyed, and in any case it was impossible to search all the rugged mountain valleys for wild fruits that might harbor the flies. Nor would insecticides prove more than locally and temporarily successful when there were so many wild fruits.

Clearly, different tactics were in order. The Hawaiian Board of Agriculture persuaded the eminent Italian entomologist Filippo Silvestri to come to Hawaii and advise them on a biological control program. Silvestri had assisted in the introduction of parasites of the gypsy moth into the eastern United States, and he was involved in the successful biological control of the white scale of mulberry in Europe. He set out to look for parasites of the medfly in West Africa and later in South Africa and Australia. He sent back several small wasps, which became established in Hawaii and rendered some measure of control of the medfly. Silvestri's report of his trip to Africa was published as a 176-page monograph by the Hawaiian Board of Agriculture in 1914.

But we cannot pass up Silvestri with so brief a mention. He was a truly renaissance figure in the field of entomology, with interests in every aspect of the science and with an especially keen eye for the small and seemingly insignificant. In the course of his long career he described two entirely new orders of insects or near-insects, the Zoraptera and the Protura, and he wrote prolifically on many other subjects, but especially on parasitic insects and biological control. His laboratory, at the Royal Agricultural College near Naples, received little financial support, but by hard work he earned an international reputation and is justly regarded as "the most remarkable entomologist of his generation." Those who knew him described him as tall and handsome, genial, and ready to help anyone who sought to tap his formidable knowledge of insects. He presided at the International Congress of Entomology held in Ithaca, New York, in 1928, and went on to give lectures at the University of Minnesota. A year before his death at the age of seventy-five, in 1949, he presented a long and important paper on biological control at an International Congress in Stockholm.

But Silvestri's life was not without controversy. He began his career as an assistant to Antonio Berlese, author of a major textbook of entomology but best known today as the inventor of the "Berlese funnel" for extracting small organisms from soil and detritus. Berlese later moved to Florence, leaving Silvestri in charge of the laboratory near Naples. Silvestri's enthusiasm for biological control was such that he believed in making use of all the natural enemies he could find, while Berlese believed that it was safer to introduce only one or two key parasites and thus minimize competition among enemies of a given pest. The disagreement between these two eminent entomologists became heated and even reached the Italian newspapers (entomologists still disagree on these matters).

In any case Silvestri was responsible for introducing several parasites of the medfly into Hawaii—and there is reason to believe that the most successful of these, a small wasp called *Opius tryoni*, might have done a better job without competition from the others. Later on, *Opius tryoni* was found also to attack the pamakani gall fly, which had been introduced to control a pestiferous weedy shrub called pamakani. In the final analysis, biological control of the medfly in Hawaii was not a huge success. Ironically, the decline of the medfly resulted not from natural enemies but from competition with an even more successful and destructive fly, which arrived in Hawaii in 1946—the Oriental fruit fly.

When medfly maggots were found in grapefruit in Orlando, Florida, in 1929, there was great alarm, and there was no thought of anything as subtle and uncertain as biological control. Rather, quick eradication of the pest from its toehold in a major citrus-growing area seemed essential. Twenty counties were found to be infested and these were promptly quarantined by road blocks manned by national guardsmen. Some 6,000 inspectors were employed to cut open fruits in the search for medfly maggots and to destroy the fruit from infested trees. After the expenditure of seven million dollars and the loss of many millions in revenue to citrus growers, the fly was in fact exterminated in less than two years. But in 1956 there was a second invasion, this time chiefly around Miami and Tampa. Once again state and federal forces were mobilized. But by this time a good deal of progress had been made in spotting and controlling infestations and it was not necessary to resort to such wholesale destruction of fruit.

Researchers in Hawaii had found that angelica seed oil could be used to trap large numbers of males. Over 50,000 plastic traps baited with this oil were placed throughout Florida to capture males and thus assist in mapping infested areas. Unfortunately the annual world production of oil of angelica was only 600 pounds, and by December of 1956 all had been used up. But researchers had been busy testing various synthetic substances, and one of these, called "trimedlure," proved highly effective and easy to produce in quantities. Like angelica oil, trimedlure attracts many more males than females, even though the evidence suggests that it mimics a pheromone produced by the males. The reason for this seems to be that male mating aggregations, or leks, form as a result of mutual attraction via this pheromone.

Once infected orchards had been plotted by trapping males, newly developed "bait sprays" could be applied to these areas. Actually, Antonio Berlese had suggested sprays baited with substances attractive to fruit flies half a century earlier, but they had had limited application prior to the 1950s. Proteins derived from yeast, bran, or soybeans were combined with an insecticide. Females require protein to develop their eggs, and they devour it eagerly, along with the poison. The mixture was applied from planes to some 750,000 acres, including even Greater Miami, where there were citrus trees in many people's back yards. In the Everglades, helicopters and swamp buggies were used to locate infestations in wild fruits. There were problems: tropical fish breeders complained of the death of valuable fish, and there was damage to paint on cars. But the program was effective, and once again the medfly was eliminated. Later local infestations in Florida in 1962 and 1981, in southern Texas in 1966, and in the Los Angeles area in 1975 were quickly snuffed out.

But in 1981 the medfly appeared in numbers in Santa Clara County, California, and began to spread to adjacent counties, threatening even the San Joaquin Valley, which grows something like a third of all United States produce. The medfly became a media event. "Fly Wars" T-shirts became popular; gift shops featured medflies imbedded in plastic fruits; the Silicon Valley developed a video game called Medfly Mania. And Governor Jerry Brown became embroiled in controversy just when he was gearing up to run for the United States Senate.

Governor Brown, mindful of the vote of conservationists and a conservationist himself, chose to adopt an approach that would do minimum damage to the environment even though his advisers urged him to support widespread aerial spraying. Several communities, in fact, passed ordinances forbidding the use of spraying from the air. The mayor of Palo Alto remarked that sentiment against the dumping of poisons from the air was "the strongest outpouring of feeling since the Vietnam War." The insecticide of choice, malathion, had been accused of being a carcinogen and causing birth defects. To forestall the use of the insecticide, the California Conservation Corps and the National Guard were mobilized to collect suspected fruit and cart it away, some 750 tons eventually being dumped into Santa Clara landfills and buried.

Recent successes with the control of screwworm in Florida and elsewhere by the release of males that had been sterilized by radiation gave hope that this might provide an environmentally sound way to reduce medfly populations. By releasing millions of sterilized males, entomologists hoped to force many of the females to mate with these males and to lay infertile eggs, which do not hatch. Fortunately, fruit flies are easy to rear on artificial media, and several laboratories were producing sterile flies in quantities. So they were released in great numbers not only in areas known to be infested but in the surrounding countryside, hoping to discourage the spread of the medfly.

But fruit growers were by no means convinced, especially since Secretary of Agriculture John Block threatened to quarantine all California produce subject to attack. Twenty-seven members of the California congressional delegation petitioned Secretary Block to override the governor and local communities. Then, in early July 1981, it was discovered that some of the "sterile" flies that had been released were not sterile at all—including some that had been released in an area not previously infested! The chief of Governor Brown's medfly program claimed that as many as 50,000 mislabeled, fertile flies may have been released and that these may have caused 95 percent of the infestation.

Remarking that the federal government had "put a gun to my head," Jerry Brown agreed on July 10 to permit use of malathion bait-sprays, applied from helicopters. Residents were assured that the sprays would do no harm to people, pets, and beneficial insects. To prove the point, B. T. Collins, director of the California Conservation Corps, drank a glassful of dilute malathion, presumably without ill effects. Some residents did object to the noisy invasion of the air space over their homes, and some complained of increases in aphids and whiteflies as a result of the destruction of beneficial insects. But on the whole the program went well enough.

By mid-August the issue was still in doubt, and the finger of blame was pointed at Jerry Brown for having delayed the start of aerial spraying. Japan, Taiwan, and South Korea, all major importers of California fruits, had threatened quarantines on a variety of fruits and vegetables. But over 100,000 traps, baited with trimedlure, registered a slow decline in medfly populations, and by December most spraying had been discontinued. When the flies failed to make a comeback the following year, those involved in the eradication program held a champagne party on October 21, 1982.

In the meantime, however, Jerry Brown's political fortunes had plummeted. One poll reported that 72 percent of the voters rated his performance as poor, citing especially his slow response to the medfly. To what extent the medfly was responsible for his loss to Pete Wilson, the former mayor of San Diego, in the congressional elections, is hard to say.

At least the media made much of it. "From Shoo-In to Scapegoat," *Time* headed a report. "Scapegoat" is perhaps the right word. The panic that prevailed in the early summer of 1981 was unfortunate (except for Brown's political rivals). The more modest efforts that Brown had initiated might have succeeded if they had continued. And he cannot be blamed that some of the "sterile" flies that were released were in fact fertile; the U.S. Department of Agriculture had handled that program. At no time was it ever determined how well adapted the medfly was for California conditions. Unlike many temperate-zone insects, the medfly is unable to undergo diapause (a prolonged, cold-resistant resting stage). At the very least, cool winter temperatures in California must greatly prolong its life cycle, such that modest efforts (such as biological control or the release of sterile males) might be able to counteract its population increases. Donald Dahlsten, of the University of California in Berkeley, remarked that the program against the medfly was conducted in a "knowledge vacuum." Paul Ehrlich, at Stanford, called aerial spraying a "serious error" that may or may not have been responsible for the ultimate decline of the medfly. One Department of Agriculture official told Ehrlich "we're in the middle of an eradication program and can't afford the luxury of research."

Although the medfly invasion of central California ended in 1981, the repercussions of the event have been more lasting. The claim that malathion is carcinogenic had been made by Melvin Reuber, a reputable toxicologist then with Litton Bionetics, which was under contract with the National Cancer Institute. But Reuber did this research on his own time and gave his results (which were not reviewed by his peers) to reporters and to the John Muir Institute, a California-based conservation foundation. Officials of the California Department of Food and Agriculture accused Reuber of scaring the people of California and having "a very negative effect on the political and public support" for malathion.

Reuber received a scathing letter of reprimand from his superiors for sending out a personal report and creating the impression that he spoke for the National Cancer Institute (which had found malathion safe). Although this letter was supposedly confidential, it was somehow leaked to lobbyists for the chemical industry, who made use of it on Capitol Hill to discredit their opponents. Reuber, now out of a job, filed suits totaling some twenty million dollars against the lobbyists and against the National Cancer Institute, claiming violation of privacy and destruction of his career. The trial is likely to be a sticky one, neither side being wholly without blame.

The tiny medfly has clearly introduced a good deal of turmoil into human affairs, including the threat to a few careers. But Jerry Brown, I am confident, will be back. And so, no doubt, will the medfly.

10

The Bee-wolf Revisited

In my 1963 book *Wasp Farm* I
included a chapter called "The Lair of the Bee-wolf," recounting
some of Niko Tinbergen's classic experiments on homing and
prey capture and my own experiences with three species that
lived around our home in Ithaca, New York. I have since moved
west, where bee-wolves abound (however unnoticed by residents and visitors!). They have provided many an exciting day
in the field, often in some of the most beautiful country on earth.

The name "bee-wolf" was first applied to a European and
African species that Fabre called *Philanthus apivorus* in his *Souvenirs Entomologiques*. This is Greek and Latin for "flower-loving
bee-eater," a thoroughly appropriate name for these insects,
since the adults feed at the nectar of flowers and capture bees to
feed their larvae. Unfortunately, Fabre's name proved incorrect,
and the species is properly called *Philanthus triangulum*. This is
the species studied by Tinbergen and his coworkers, more recently by Terko Simonthomas, formerly of the Pharmacological
Laboratory of the University of Amsterdam, in The Netherlands, assisted by his wife Miek.

The bee-wolf of Europe is unpopular with the beekeepers,
since it preys exclusively on honey bees. Simonthomas reports
that a female may catch as many as ten honey bees in a day. A
population of 3,000 wasps is not unusual, he reports, and such a
population may capture 30,000 honey bees in one day—the
number in an average hive. Because of his familiarity with the
bee-wolf, Simonthomas was invited by the Egyptian govern-

ment, in 1978, to study a serious problem with bee-wolves in the Dakhla oasis in central Egypt. He had an ulterior motive in accepting, since his laboratory was involved in the pharmacology of animal venoms, and here was an opportunity to collect large numbers of wasps and extract their stings and associated glands.

The Dakhla oasis had been an important honey-producing area for some time, using an aggressive strain of bees known as Egyptian bees. In 1960 the government decided to replace these bees with a race that was more docile and easier for beekeepers to handle: the Carniolan bees popular in Europe. Unfortunately, the bee-wolves also appear to have found them easier to handle, for they built up huge populations at the expense of these more defenseless bees. By 1972, honey production was nearly down to zero, and there was only slight recovery in the following years despite efforts to control the bee-wolves. In 1976 the government offered a bounty of one piaster for each bee-wolf destroyed. This resulted in the capture of 24,000 bee-wolves in two months. But the following year they were nearly as abundant as before.

Bee-wolves prefer to hunt at flowers, but during winter at the Dakhla oasis there are few flowers in bloom, and the wasps hunt at the hives. They often land on the flight board at the entrance and seem to watch the bees going in and out; at an appropriate moment they tangle with a returning forager, sting it, and fly off to the nest. If there are guard bees at the entrance, the wasps usually don't succeed, and if several guards attack them they may be killed. As soon as spring flowers appear, the bee-wolves move to them, where they are less often noticed by beekeepers but are just as destructive, probably more so, since individual bees collecting nectar or pollen are quite defenseless.

Terko Simonthomas pointed out that strong colonies of honey bees usually have numerous guard bees and are thus able to defend themselves well against the wasps. So one of his suggestions was to combine weaker colonies to form stronger ones. Another, more radical suggestion was to remove all bees from the oasis for one year, thus "starving out" the bee-wolves. This was not a popular suggestion, since the bees were needed to pollinate crops grown in the oasis, and in fact was never followed.

Although there are more than thirty species of *Philanthus* in North America, none of them prey upon honey bees to any great extent. Honey bees are, of course, native to Eurasia and Africa, and even though they have been on this continent for three hundred years or more, none of our bee-wolves have come to use them in any great numbers. There are records of four species using honey bees, and presumably if one of these became established near an apiary it might cause a problem. But all of our species seem to prefer wild bees of one kind or another, or sometimes wasps, and our studies have therefore been conducted on minimal budgets. Since the importance of wild bees in plant pollination is usually overlooked, it is hard to convince anyone that their predators pose enough threat to society that their study is worthy of support. But it has been great fun, and has supplied several students with some fine doctoral dissertations.

North America is home to the largest bee-wolf in the world, about an inch long: *Philanthus bicinctus*, literally the "twice-banded" *Philanthus*. The name is appropriate enough, as the females have an orange and a yellow band at the base of their abdomen. We prefer to call *bicinctus* the bumblebee-wolf, since it is a formidable enemy of bumble bees. One of the pleasures of studying these wasps is their predilection for nesting in scenic places. My students and I first encountered bumblebee-wolves in Yellowstone National Park, where they inhabited a hillside overlooking some hot springs as well as the junction of the pristine Snake and Lewis Rivers. Later we studied them at Great Sand Dunes National Monument in southern Colorado, where massive hills of sand are piled against the foothills of the Sangre de Cristo Range. Still later and closer to home, we found them nesting near Chimney Rock, a great pillar of sandstone on the Colorado-Wyoming border. All three sites are at approximately the same altitude, about 8,000 feet, where one can count on fresh mountain air and the bluest of skies—except for sudden thundershowers, which send the wasps to shelter (unfortunately the nearest human shelters are in each case some distance away!).

At Chimney Rock, in late summer, the slopes are a mass of flowers, especially wild buckwheat and a variety of composites such as yellow snakeweed and purple asters. So there are plenty of bees. Four kinds of bee-wolves nest there, and we were interested to see how four closely related species were able to thrive in the same habitat. *Philanthus bicinctus,* true to form, preyed only on bumble bees. A slightly smaller species, called *basilaris,* used an occasional bumble bee but mostly preyed on a variety of medium-sized bees and wasps. A third species, *inversus,* although about the same size as *basilaris,* proved to be a specialist on males of one kind of green sweat bee. The final species *barbiger,* was a tiny wasp, preying only on quite small bees and wasps. So together the four species are poised to exploit the great numbers of flower-visiting bees and wasps each all-too-short summer, but without competing seriously for the same kinds.

At Great Sand Dunes, five species of bee-wolves nest, with a similar tendency to share the available prey. The bumblebee-wolves there have been studied for several years by Darryl Gwynne, whom we met in an earlier chapter as a devotee of Mormon crickets. For at least ten years, about 200 individuals of each sex have occupied a sloping, sandy bank sparsely clothed with grasses, yuccas, and cacti. In an average year, each female makes a nest of about ten cells, each cell provided with about five paralyzed bumble bees. Thus the local bumble bee population is depleted by about 10,000 each year. In 1978 bumble bee populations were unexplainably low throughout the Rocky Mountains, and that year over half the prey captured were large leafcutter bees and certain kinds of large digger wasps. So bumblebee-wolves are sufficiently versatile to use other kinds of prey if need be, and so far as we know their larvae do perfectly well on any kind of bee or wasp.

Darryl Gwynne's major research was not on the females, however, but on the remarkable behavior of the males. These are striking wasps, banded with yellow in a pattern quite different from the females. Even more striking is their behavior in the territories they establish among the nests of the females. The territories are roughly a meter in diameter, and the best (that is, most frequently occupied) territories are located where there are the most nests. Gwynne made several imitation nests apart from the major nesting area. Sure enough, males set up territories at the spot, responding to the holes and piles of sand he had provided and demonstrating that they are visually attracted to nests.

Males spend the night in shallow burrows in the nesting area. Then, about nine o'clock, they emerge, fly about a bit, and begin to fight for the best territories. Once a male has succeeded in establishing himself, he begins to lay down a sex attractant pheromone on grass stems and other vegetation surrounding the territory. The pheromone is produced from large glands in his head, applied by paired brushes just above his mandibles, and spread on the plant with dense hairs on his abdomen. He first lands on a stem, then walks up a short distance, turns around, and walks down again, all the while pressing his head and dragging his abdomen against the stem, so that his body has a broadly V-shaped posture. This is repeated many times, especially in the early part of the day. The pheromone evidently serves to attract females; at least, all the matings we saw appeared to have been initiated on the territories.

Territories are vigorously defended against other males, usually by butting them in the air or grappling with them. Insects that resemble the males somewhat (that is, any banded insects of about the same size) are approached and occasionally butted, the *bicinctus* males apparently mistaking them for members of their own species. Even passing butterflies and birds may be pursued a short distance. But the strongest reaction is to females of their own species, with which they frequently engage in "wrestling matches" in the effort to make contact with their genitalia. However, since females only mate once, and afterward reject all suitors, most of the male's efforts are in vain.

Even though there were often around 200 males in the area (Darryl Gwynne tried to mark all of them, using numbered discs glued to their backs), there were never more than forty to eighty territories, and never more than twenty to forty-five occupied in one day. Where were the rest of the males? Clearly they had been excluded from territories by interactions with other males. Some of them established territories away from the nesting area, in nearby fields of flowers where females often hunted for bees and where scattered nests occurred. Others were not seen to be territorial at all, and we assume they simply "took their chances" on meeting a female somewhere else. Generally speaking, it was the smaller males that failed to become territorial, as might be expected from the nature of the fights that occurred during territory establishment. Smaller males apparently adopt a "loser strategy," making the best of it elsewhere.

If larger males are more successful in mating, why doesn't natural selection produce wasps of increasingly large size? The answer is, of course, that there are other constraints on size. It would take more prey, or larger prey, to feed a larva that will produce a larger wasp. There are no larger bees available than bumble bees, and there is a limit as to how many a female can catch in a day, or in her brief lifetime. That some individuals are larger than others is a consequence of variation in the amount of prey in the cell, which in turn may reflect varying success in hunting or the fact that some of the prey may be consumed by mold, nematodes, or parasites, so that the larva is short on rations.

A second species of *Philanthus* occurring at Great Sand Dunes, *basilaris*, became the special province of another former student of mine, Kevin O'Neill. As at Chimney Rock, *basilaris* takes smaller prey than *bicinctus*, so the two don't compete for prey even though they nest in the same general area. Actually, female *basilaris* don't nest in aggregations like most bee-wolves; their nests are scattered individually over a wide area. So, even though males are territorial in much the manner of *bicinctus* males, it would hardly do for them to establish territories among these scattered nests. In fact, they do something quite different: they cluster their territories at specific sites.

Several males (sometimes fifty or more) establish closely adjacent territories, and females fly to these sites, where most mating occurs. These aggregations, properly called "leks," are comparable to those of medflies, discussed in the previous chapter. Evidently the females fly upwind, responding to the pheromone the males have laid down. The leks are not established in places where there are especially large numbers of flowers, so the females are evidently not attracted there for any reason other than the presence of the males. The odd thing is that the same

A male bee-wolf, poised and ready to respond to insects that traverse his territory.

P. EADES

lekking sites are used year after year, even though each year a different generation of wasps is involved. One of the sites, a thoroughly nondescript spot not far from the Great Sand Dunes visitors' center, was occupied by ten to twenty males each of five successive years. We intend to watch the site for as many more years as we can. It is good to have an excuse to visit such a beautiful spot!

When a female enters a lek there is a great furor among the males, who contest vigorously for the opportunity to mate with her. Since only virgin females mate, we at first assumed that only virgins would fly to a lek. Not so! Females that are looking for prey to provision their nests also fly to leks—and seize and sting a suitor to carry to the nest as food for her larva. We saw this several times, and on one occasion found a nest cell that contained two paralyzed male *basilaris* which the larva was about to consume. Cannibalism is not rare in some other insects (such as robber flies), but the capture of males to feed the offspring is unique. Since we also found a male of *bicinctus,* the bumblebee-wolf, in a nest cell of *basilaris,* we facetiously came to call the latter species the bumblebee-wolf-wolf.

We had now studied a species *(bicinctus)* that nested in a moderately dense aggregation and in which males established territories among the nests; and a second species *(basilaris)* in which the nests were scattered widely and in which the males formed leks. We knew that there were species in which the females formed very dense aggregations, their nest entrances closely adjacent to one another. What strategy would the males use? Territories among the nests would be difficult to maintain because of the severe competition for the limited space; and leks would seem inappropriate. Such a species was available to us, *Philanthus zebratus,* which might be called the zebra-striped bee-wolf. Study of this wasp took us to another delightful place to spend a summer, Grand Teton National Park, Wyoming. Fortunately, there is a good Biological Station in the park, maintained by the University of Wyoming and the National Park Service, and we have been there so often (beginning in 1961) that I have lost track of the times. An aggregation of *Philanthus zebratus* has been active at Deadman's Bar, close beside the Snake River, for at least twenty years. Despite the temptation to study some of the many other wasps occurring in the park (to say nothing of the temptations of backpacking and fishing), Kevin O'Neill and I have found time to observe these wasps at some length. The males do, indeed, have a mating strategy quite different from other species and uniquely suited to the very dense nesting aggregations of that species.

What happens is this. Females, after digging their nest, make a very high, circling orientation flight, up to about three meters above the nesting site. Then, when they return to the nest after taking nectar or catching a bee, they return at this height before descending to the nest. Males as usual spend the night in short burrows in the soil. Then, in mid-morning, when the females are making their first flights from their nests, they ascend to a height of about three meters and fly about at that altitude for a few seconds before returning to the ground, then ascending again a short time later. Since most of the males do this, 100 or more, at any one time there is layer of males in the air at about the same height, the height to which the females rise or begin their descent to the nest. Contact between the sexes occurs at this height, and the pair then descend to a plant to complete their mating.

But *zebratus* males have not abandoned territoriality. Spaced along the edge of the nesting area, on many days, we found several territories, defended and marked with pheromone in the usual *Philanthus* manner. But oddly, the territorial males average smaller than those participating in the "high flight." Apparently territoriality is a "loser" rather than a "winner" strategy in this species. A few males were found to be territorial on one day and to patrol in the air another day, so apparently males have the option of doing one or the other. In a second, smaller and less concentrated nesting aggregation just north of the park, we found nothing but territorial males, so apparently the "high flight" behavior is elicited only in situations where there are many nests crowded into a small area.

Since we have studied more than half of the approximately thirty species of *Philanthus* occurring in North America, it would be easy to prolong this chapter indefinitely. Kevin O'Neill and I hope someday to write a book on the subject, so we shall hold back some of our secrets. It does appear that mating strategies vary depending upon nest density, among other things, and that larger males are able to undertake, through aggressive interactions, whatever tactic is most successful in that situation. What, in turn, determines nest density? Presumably the amount of suitable soil available at the site is important, and species differ in the type of soil they prefer. Some kinds of soil are available in broad expanses (for example sand dunes), others only in small patches here and there. The availability of suitable prey also plays a role. How the males perform seems very much dependent upon how the females respond to these variables.

The life history of any species is an integrated whole, no part of which makes sense by itself. That is why there is still need for persons, amateur or professional, willing and eager to spend endless hours in the sun, wind, and rain gathering bits and pieces of data. Knowledge is built of bits and pieces, some large and some seemingly inconsequential. All of them fit together into a total view of the world we live in, a view derived from myriads of observations and made meaningful by concepts that wrap these observations into bundles that can be appropriately labeled and called for when needed for utility or for refreshment.

There is much to be said for making discovery a way of life, however limited one's resources. In Jacob Bronowski's words "science is nothing else than the search to discover unity in the wild variety of nature." It does not pay to worry about the usefulness of a discovery; all we know of the world is the result of discovery pure and simple, and all of technology is the stepchild of science.

Now and then, on cold winter days, I like to mull over these high-sounding ideas. But perhaps I am merely rationalizing my summers spent with a camera, notebook, and tape recorder in a place where the buzz of insects fills the air and bee-wolves are simply being bee-wolves, which is enough.

11

Marsh Flies

During the last few months of World War II, I was assigned to an army base hospital as a parasitologist. The U.S. forces had just reoccupied the Philippines, with considerable loss of life and the loss of many man-days as a result of disease. During the recapture of Leyte, the troops had to cross farms and wade through rice paddies, where they became infected with a variety of parasitic worms. It was our job, after the casualties arrived in our hospital, to determine the level of infection by counting the worm eggs in stool samples. In a grim and odorous way, it was rather fun.

Each morning our laboratory was presented with a huge pile of stool samples, each in a neat cardboard container. After eliminating most of the fecal material by certain techniques, we were left with particulate matter among which were the eggs of hookworms, blood flukes, and others. My favorite was a curious brown, bottle-shaped egg, that of a worm with the euphonious name *Trichuris trichiura*, the human whipworm. But whipworms usually do not produce serious symptoms; we were mainly concerned with blood flukes, which often produce painfully enlarged livers and spleens and an otherwise emaciated condition. There were some very sick people in our hospital. Now and then we assisted the M.D.s when they made spleen punctures on advanced cases. That part was less fun.

Infection with blood flukes is called schistosomiasis, or snail fever, and the fluke in this case was *Schistosoma japonica*, a serious pest in many parts of the Orient. Some 1,700 U.S. troops succumbed to infection on Leyte and were put out of commission. *Schistosoma* literally means "split body," and refers to the fact that the male worm has a longitudinal slit in its body in which the more slender and elongated female rests, "safe in the arms of her spouse," as one parasitology text puts it. The worms occur in pairs in the blood stream, where the female lays her eggs. The eggs eventually make their way to the intestine and thence to the feces. The problems are caused by the accumulation of eggs in the tissues and the swelling and inflammation of tissues as a reaction to the eggs. Lesions may occur in the intestine, liver, spleen, and even in the brain. While the result is rarely fatal, perhaps as many as 200 million people in warmer parts of the globe function at less than full capacity as a result of infection by *Schistosoma japonica* and two other, related species of blood flukes.

The life cycle of these worms is fairly bewildering. The eggs hatch in water and produce tiny, swimming organisms called miracidia. In our laboratory we often placed eggs in water and watched the miracidia propel themselves about a drop of water under a microscope. In nature, the miracidia must find a snail of particular kind and enter it; here, after a few weeks, their bodies fill up with tailed offspring, a bit like microscopic tadpoles, produced in the absence of sexual reproduction. These are called cercaria, and they escape from the snail and swim about in the water, ready to penetrate the skin of an animal that has occasion to bathe or wade in the water. *Schistosoma japonica* infects cattle, goats, pigs, dogs, cats, and other animals as well as humans, and since manure is used in rice paddies and for other crops in the Orient, there are infective cercaria almost everywhere. The life cycle is, however, rather precarious. The miracidia must find a snail of proper species within a few hours, and the cercaria must find a vertebrate host within two or three days; otherwise they die, as many must do in nature.

While schistosomiasis is not a threat in temperate parts of the globe, there are schistosomes that infect ducks and have temperate-zone snails as alternate hosts. The cercaria occur in

A marsh fly's elegant lines belie its role as a vicious predator of snails.

P. EADES

lakes in many parts of the world, where they sometimes penetrate the skin of bathers, causing "swimmer's itch." The cerceria ordinarily die in the skin of humans and cause only a local rash, which disappears in a few days.

Blood flukes are not the only parasitic worms that use snails as alternate hosts. A larger, leaf-shaped worm, called the liver fluke, occurs in many parts of the world and develops in the bile ducts of cattle, sheep, goats, and humans, causing a weakened condition and sometimes death resulting from various complications. The cercaria of the liver fluke cannot bore through the skin but must be swallowed. In India and in parts of South America, particularly, a high percentage of cattle bear liver flukes, and they are not at all uncommon in humans. Neither blood nor liver flukes can be transmitted directly, from one person to another. In every case the parasite must spend part of its life cycle in a snail before becoming infective to a warm-blooded animal.

What has all this to do with marsh flies, or to insects of any kind? I certainly saw no connection when I was counting worm eggs back in 1945. Since schistosomiasis is difficult to treat—some treatments are worse than the disease—attention has centered on the control of snails. And it happens that marsh flies attack snails and constitute the only major group of organisms that feed almost exclusively on snails. This fact was, however, not appreciated until some time after World War II. Through a fortuitous series of events, I had an opportunity to watch the development of knowledge of the biology of marsh flies and their possible use in the control of snails. In fact, several of the participants are good friends and from time to time have permitted me to have intimate glimpses of a group of flies I had scarcely known existed up to that time.

Very little was known about marsh flies until fairly recently. In his influential text *An Introduction to Entomology* (1924), John Henry Comstock remarked only that these flies "are usually found in moist situations, as along the banks of streams. The larvae are aquatic." A few had been reared from snails by European workers in the early part of the century, but nothing had been learned about the nature of the association.

The story really begins in 1950. During that year Clifford O.

Berg, then at Ohio Wesleyan University but soon to move to Cornell University, was studying mosquitoes in Alaska as consultant to the Arctic Health Research Center. Dipping for mosquito larvae in pools in the tundra, he picked up some other larvae he did not recognize. He put them in a dish in the laboratory and, not knowing what they fed on, casually dropped in a snail. Within a few hours the larvae had gorged themselves, leaving the snail shell "as clean inside as a museum specimen." When the adults were produced they turned out to be marsh flies.

Over the next few years Berg reared several species from snails in Alaska and in Michigan and Ohio. In 1953 he was emboldened to suggest that all members of the family of marsh flies (Sciomyzidae) might be found to develop in mollusks and that they might prove useful in controlling undesirable snails. At Cornell, Berg had a series of brilliant graduate students who worked closely with him, and by 1971 he was able to say that his group had investigated the life cycles of 200 species of marsh flies—more than a third of all the known species in this family. He was also able to state that one species he and his students had found in Central America and shipped to Hawaii was well established there and attacking the snail hosts of the giant liver fluke, a serious pest of cattle on the islands.

Marsh flies are rather beautiful little creatures, more slender than a house fly and often adorned with a complex color pattern on the body and wings. They occur throughout the world, from the arctic to the tropics, and are often abundant where snails occur. The adult flies feed on flowers, dead insects, or sometimes on the mucus of snails. They are relatively long-lived for flies, some having lived in the laboratory for more than a year when supplied with adequate food. Most larvae are relatively host-specific, attacking only snails of a certain kind. A few attack slugs, which is not surprising since slugs are essentially snails without shells. They are, however, terrestrial (like some snails), so it is not correct to say that marsh fly larvae are always aquatic.

Study of marsh flies has yielded an endless series of surprises for Clifford Berg and his students, so diverse are the lifestyles of these insects. This diversity makes sense if considered in a framework of evolution. The ancestral marsh flies may have been scavengers, feeding on dead, decaying snails, as a few still do (the scavenging habit is very widespread among flies). A much larger number of marsh flies attack living snails as predators. Their larvae lacerate the tissues with their mouth hooks, feeding until they are satiated. Then, after a time, they may attack another snail. Some have been found to kill and devour more than a dozen snails during their larval lifetime. Sometimes, in the laboratory, several larvae will feed together on a snail without attacking one another. In these two groups, the scavengers and the predators, the adults generally lay their eggs on vegetation in wet places, and the larvae find snails under their own power.

There are other, more advanced species of marsh flies in which the female selects a particular kind of snail and lays her eggs directly on it. The larvae then enter the snail and develop slowly within it, feeding first on nonvital tissues and finally killing the snail only when they are nearly mature. Only one larva develops per snail, apparently because the first larva to enter destroys any that appear later. In most cases the larva forms its pupa within the snail shell. Such marsh flies are termed "parasitoids"—they are not true parasites since they kill their host, and not predators since they feed slowly and devour a single host.

There are enough intermediates between these major types (scavenger, predator, parasitoid) to reinforce the idea that existing marsh flies represent a sampling of successive stages in evolution. There are also a number of marsh flies of quite specialized behavior. A few, for example, lay their eggs on the egg masses of snails, and their larvae consume the snail eggs. The species that attack slugs are able to follow the slime trails of their hosts; having found a slug, they paralyze it with a powerful nerve toxin before they begin feeding. None of the snail-feeders are known to use such a toxin.

Determining the normal host of a given marsh fly involves capturing adult females, getting them to lay eggs in the laboratory, and then giving the newly hatched larvae a choice of snails available in the natural environment. Clifford Berg and his collaborator Stuart Neff were interested in finding the normal host of a species confined to salt marshes along the eastern coast of

North America. They had no trouble obtaining newly hatched larvae, but the larvae showed no interest in either of two common salt marsh snails. They did feed intermittently on several freshwater snails, but it was clear that these were not normal hosts. Eventually, Berg and Neff found larvae in debris along the waterline where common saltwater periwinkles were abundant. These seemed an unlikely host, since they are operculate snails, having a tough cap that seals them off inside their shells. Other marsh flies attack nonoperculate snails, and it is known that in some cases if they attack operculate kinds the larvae are effectively squeezed to death when the operculum closes. But when newly hatched larvae of the salt marsh fly were offered periwinkles, they attacked them vigorously, apparently gaining entry along the curled edges of the cap.

Discovery that a few marsh flies attack clams was equally unexpected. Apparently two quite unrelated groups have independently acquired a taste for clams, in both cases for tiny, freshwater fingernail clams. In some cases a single larva will devour the soft tissue of as many as fifteen or twenty clams.

This is only the barest outline of a few of the recent discoveries concerning marsh flies. Since most species can be reared fairly easily in the laboratory, the group has provided a wonderful series of variations on a single theme: mollusk-eating. It is not often that one can learn so much detail about the majority of members of a single group of animals. As Berg and his collaborators have pointed out, speculation on the evolution of a group more clearly approximates reality as the percentage of studied species increases.

What of the use of marsh flies in biological control? I have already mentioned that a Central American species has been successfully introduced to Hawaii for control of the giant liver fluke. In 1967 a second species was brought from Japan to Hawaii in the effort to provide further control of this fluke. These flies are well established on all the major islands and there has been a decrease in the number of infections of liver flukes in cattle in recent years, although this cannot be attributed to the flies with certainty. Fortunately, neither of the introduced marsh flies attacks the beautiful tree snails of Hawaii or the predatory snails that had been introduced earlier to control the giant African snail, a serious agricultural pest.

With respect to blood flukes, the problems are formidable, as these worms and their snail hosts occur over vast areas of the tropics. In the laboratory, the snail hosts of several species of schistosomes are vulnerable to attacks by marsh flies. Unfortunately, the snail hosts of *Schistosoma japonica* have tough opercula which the larvae cannot penetrate, so this species may remain immune to this type of biological control. It seems probable that reliance will have to continue to be placed on molluscicides, even though these potent chemicals often kill fish and many other kinds of aquatic organisms. As increased damming and new irrigation projects in underdeveloped countries continue to provide new habitats for snails, it would certainly seem appropriate to continue to investigate the potential of marsh flies. They are not likely to provide a panacea for schistosomiasis, but even if they were to play a minor role it would be worth the effort.

Clifford Berg's discovery of marsh flies as natural enemies of snails while he was dipping for mosquito larvae is a prime example of serendipity—the gift for making fortunate discoveries accidently. There are so many examples—Thomas Hunt Morgan's discovery of the white-eyed mutant *Drosophila* and Sir Alexander Fleming's discovery of penicillin, to mention only two. I tend to turn purple when I read comments such as the following: "basic work in general should arise from the need to answer questions coming from more applied work rather than preceding such work" (extracted from a recent review of a research institution). So much has been learned by those who were merely curious; and so much has resulted from following up on accidental discoveries! Who can say what obscure group of organisms, what seemingly insignificant bits of information may prove useful to us? All are useful in the sense that they allow us to exercise the most unique attribute of our species: the ability to understand (to a degree) the world we live in.

12

The Milkweed Bug

Most insects that have become well known have achieved their fame (or notoriety) because they are "good" or "bad" by our reckoning. Everyone will agree that lady beetles are nice to have around; they eat aphids, mealybugs, and other pestiferous creatures. What insects are "bad" and how bad they are is often a matter of opinion. Being an entomologist, I tend to admire insects that have been the objects of much valuable research; and I must honestly admit that much of the support for entomological research results from the damage insects sometimes do. But if I were a cotton farmer I would not speak kindly of the boll weevil, and if I were a California fruitgrower I would have no love for the medfly!

Besides "good and bad" insects there are some that have achieved fame simply because they are easy to keep in the laboratory and so provide systems that are convenient to tinker with on cold winter days. *Drosophila* provides the best example. Who would have supposed that "stupid little saprophyte" (as Professor W. M. Wheeler used to call it) would teach us much of what we know about genetics? In nature, the flies merely swarm about overripe fruit and make up for the minor nuisance they cause by helping in the decomposition of decaying organic matter. But when it was found that they could easily be reared through many generations in milk bottles, they became the focus of research out of all proportion to their ecological significance.

The milkweed bug is another easy-to-rear insect that dozens

of laboratories have found convenient to study. It was as late as 1934 that Floyd Andre, at Iowa State University, discovered that the bugs can be cultured indefinitely as long as they have a warm place to live, plenty of milkweed seeds, and some water. Since that time a great deal of attention has been devoted to this insect of no importance to humanity (milkweed being of interest chiefly to addicts of natural foods and to poets and photographers). But researchers hope that some of the things that have been learned about the milkweed bug can be extrapolated to other animals, and of course they are right.

Milkweed bugs have sucking beaks consisting of several piercing stylets enclosed in a sheathlike labium. Thus they are placed in the order Hemiptera and can properly be called "bugs" (entomologists frown on the use of this word for insects of other orders, though it doesn't do them much good). There are, in fact, two kinds of milkweed bugs in North America. The best-studied species, and the one that is the hero of this chapter, is properly called the large milkweed bug *(Oncopeltus fasciatus).* The other is (predictably) called the small milkweed bug *(Lygaeus kalmii*—named for Peter Kalm, whom we'll meet in Chapter 15).

Both large and small milkweed bugs are attractive insects, liberally splashed with orange, like several other insects that feed on milkweed, such as the milkweed longhorned beetle and the monarch butterfly. In the case of the monarch it is known that the caterpillars take up digitalis-like substances from the plants and sequester the chemicals through to the adult stage. Birds feeding on the butterflies are made ill and learn to associate the orange color pattern with toxic properties.

G. G. E. Scudder and his associates at the University of British Columbia have shown that both species of milkweed bugs also take up these digitalis-like substances (cardiac glycosides) from the milkweeds. The substances are absorbed through the wall of the midgut and become concentrated beneath the epidermis and in special glands along the sides of the body. There is evidence that birds and lizards quickly learn to associate the bright colors of milkweed bugs with the noxious properties. Alan Gelperin, at Princeton University, kept several praying mantids in his laboratory, feeding them with blow flies and with milkweed bugs. He found that on their first encounter with a milkweed bug, mantids strike, capture, taste, and then discard the bug, sometimes vomiting afterward. On subsequent encounters they discard it without tasting, and eventually they simply stare at the bug and refrain from striking at all. So evidently brilliant orange-and-black patterns are sometimes effective against insect as well as vertebrate predators.

Scudder found that milkweed longhorned beetles retain only small quantities of toxins in their blood, so presumably they have evolved orange colors as mimics of more distasteful species. Grasshoppers and other insects that feed on milkweeds do not absorb any of the glycosides, so they remain perfectly palatable to predators.

Milkweed bugs, like other true bugs, also have scent glands, loaded with repugnant chemicals similar to those of stink bugs. Curiously, the glands of adult males are larger than those of the females and produce a different medley of chemicals. Among these are certain acetates that have a rather pleasant smell, so that a pure culture of male bugs has a good deal nicer smell than one of females. Perhaps these acetates have nothing to do with repellence but something to do with attracting females. There seems no other good reason why males should have larger glands with different odorous products.

The most serious problem in rearing milkweed bugs has been the collection of enough milkweed seeds to keep them through the winter. The time-honored method has been to gather one's students about oneself in the fall, lecture them on the wonders of scientific research, and then send them off with burlap bags to collect as many milkweed pods as they can. That part is fun; but when they are faced later on with separating the seeds from the down, their enthusiasm is likely to become somewhat dimmed.

During World War II it was possible to buy milkweed seeds from a company in Petoskey, Michigan. This company was engaged in the manufacture of life preservers, which are normally stuffed with kapok, the product of an East Asian tree. But when supplies were shut off during the war, milkweed down was employed, yielding seeds as a byproduct. After the war, unfortunately, entomologists were once again left to their own devices. Remembering the revolution brought about by Eli Whitney's invention of the cotton gin, they began to find ways of separating the seeds from the down by machine. Certain types of vacuum cleaners worked pretty well, collecting the seeds in the dirt pan while the down clung to the bag. For large-scale collection of seeds, a threshing machine for grain could be used, provided certain adjustments were made. In this way one investigator obtained sixteen and one-half bushels of milkweed seeds from thirty-eight bags of pods. Enough to feed quite a few bugs!

It is not surprising that various people asked whether the bugs could be reared on something easier to obtain than milkweed seeds. Peanuts were tried, and the bugs would feed but grew slowly and laid fewer eggs. Other foods that were tried included walnuts, almonds, pecans, sunflower seeds, raisins, kidney beans, rolled oats, peanut butter, and coconut; but in every case the bugs didn't do nearly as well as on their normal food. The best alternative proved to be watermelon seeds, provided the seed coat was cracked. It is surprising that any of these foods worked at all, since in nature the bugs are restricted to various species of milkweed and the related dogbane family.

Stanley Beck and his associates at the University of Wisconsin found that milkweed bugs will accept an artificial diet, consisting of glucose, starch, casein, yeast, corn oil, cholesterol, and water. However, they feed and grow very slowly unless the diet is enclosed in the husks of milkweed seeds. Since they won't feed through a simple membrane enclosing the diet, it was concluded that something in the chemistry of the husks served to stimulate feeding. When the bugs' antennae are cut off or covered with shellac, they feed very little, suggesting that the antennae contain the sense organs that perceive the identifying chemicals.

Beck and his associates also showed that a milkweed bug takes from 200 to 400 minutes of feeding to deplete the food in a milkweed seed; in twenty-four hours a nearly full-grown bug will consume the contents of two or three seeds. From time to time the bugs "take a break" to obtain water. When feeding, they produce two kinds of saliva, each from different parts of their salivary glands. One kind hardens to form a sheath around the stylets, presumably serving to prevent leakage, while the other flows into the tissues and contains the enzymes that start digestion. Then the food is drawn through the stylets and into the digestive tract by means of a muscular pump in the pharynx. In this respect the milkweed bug doesn't differ much from other sucking insects. What is surprising is that an essentially dry source of food can be sucked up through very fine, threadlike stylets—rather as if one tried to eat a biscuit through a straw. Of course, this is made possible by the frequent uptake of water and the production of copious amounts of watery saliva. If you forget to supply your bug culture with plenty of water, you soon have a bunch of corpses on your hands. In nature, the bugs feed not only on seeds but on dew and plant sap; before the seeds are ripe, they feed primarily on the buds and young leaves.

It would be impossible to review all of the hundreds of articles that have been written on the milkweed bug—mostly by people in the northern tier of the United States, where entomologists tend to become lonely for live insects during the long winters. At Cornell University, Ferdinand Butt made a detailed study of the embryology of the milkweed bug in 1949, and his student Philip Bonhag, along with Bonhag's student James Wick, followed up with a monograph on the functional anatomy of the reproductive systems in 1953. These may seem like esoteric subjects, but Bonhag and Wick's superb drawings of the interlocking genitalia are something to behold. The legendary Rube Goldberg never devised anything more elaborate. The male has a remarkably long, coiled penis which he inserts deeply into the female's abdomen, all the way to her spermatheca (sperm storage pouch). The penis is extended by fluid contained in a musculated erection fluid reservoir. In order to study the functioning of the various parts, Bonhag and Wick placed copulating pairs in the refrigerator, then after they were well cooled imbedded them on their sides in paraffin and covered them with cold salt solution. As they gradually revived, the working of the various parts could be observed and pieces could be snipped out and prepared for sectioning and staining, for study of the finer structure.

Philip Bonhag and I were classmates at Cornell and later colleagues at Kansas State University. The course he gave on insect structure at Kansas State, and later at other universities, was a rich experience for students. A dead insect, dissected and pinned out in a tray, suddenly became a living thing, all of its parts significant in terms of its lifestyle. Insect anatomy is traditionally the driest of subjects, but under Bonhag's guidance "cutting up an insect" (with infinite care, of course) became the ultimate in discovery.

It was Phil Bonhag's philosophy (as he often ruminated over a glass of orange juice) that one should not spend one's life doing much the same thing (as so many of us do!) but should continually broaden one's horizons. He went on to study the chemistry of egg formation in the milkweed bug and later expanded this to a review of egg development in all insects. Had he been granted a normal life span, he would undoubtedly have taught us much about still other complex life processes. As it is, there is a bit of him in all of us who knew him.

Milkweed bugs spend a great deal of time mating. Each copulation lasts from less than an hour up to ten hours, and females must mate repeatedly for maximum egg production. Evidently the mere presence of males enhances the development of the female's ovaries. When females are reared in isolation from males, only about half of them show full development of their ovaries by seven days after they reach adulthood. But when males are present, but prevented from mating, 76 percent of the females show full development of their ovaries in the same time span. In many animals, the presence of the opposite sex speeds and fine tunes sexual development. If a female milkweed bug has no access to a mate, it is to her advantage to use her resources to prolong her life until such time as a male appears. After mating, all females show full ovarian development. Even mating with castrated males enhances egg production. (Since insect sex hormones are not produced by the reproductive organs, as they are in humans and other vertebrates, but in the head region, castrated males are as "sexy" as normal males, but unable to fertilize the female's eggs.)

Frequent mating does, however, shorten their lives. Virgin females live an average of 101 days, mated females only 45; virgin males live an average of 132 days, while those with ready

---

Milkweed bugs advertise their toxic properties with bright orange and black coloration.

P. EADES

access to females only 61 days. These findings added spice to long winters in the laboratory of Walter Lener, at Nassau Community College, New York. The milkweed bug is not unique in this respect, however. Virgins live longer than mated individuals in cockroaches, carpet beetles, and mosquitoes, and probably many other insects. This is highly adaptive, since adult insects are specialists in reproduction, and it is to the advantage of every individual to survive as long as possible when mates are not immediately available. Insects do not enjoy the long postreproductive lives that we do; once they have mated and laid their eggs, their lives are fulfilled.

In nature, milkweed bugs usually occur in small groups. Interactions between group members (aside from mating) appear minimal, and it may be that group living has evolved primarily to enhance the effectiveness of their warning coloration. Dorothy Feir, of St. Louis University, found that bugs reared alone (in laboratory cages five by nine centimeters in size) averaged smaller and had a lower percent survival than those reared in groups of from five to twenty-five. So, evidently, subtle mechanisms have evolved that cause them to favor group living. However, when 50 to 100 individuals were reared in chambers of the same size, survival was much lower and the average weight of those that survived was lower than normal—even though they had plenty of food and water. It appears that extreme crowding is even more unfavorable than solitary living. In nature, such crowding would doubtless result in destruction of their food source.

Dorothy Feir has also studied the constituents of the blood of milkweed bugs, which among other things contain a substance that causes human red blood cells to clump. It had earlier been found that some kinds of bacteria, including *Staphylococcus*, were destroyed by certain elements in the blood of this insect. Other substances isolated from insects have been shown to have antibiotic or other effects on human systems, but so far little has been done to exploit their possible uses in medicine.

Some of the most interesting recent research on milkweed bugs relates to their migratory behavior. Unlike the small milkweed bug, which overwinters in the northern states, the large milkweed bug is a member of a tropical group of insects that lack the ability to hibernate. In North America, the bugs spend the winter in the deep South and in Mexico and the West Indies. In spring, they begin to make their appearance farther north. Each succeeding generation proceeds northward by a series of steps, winged adults apparently taking advantage of warm, southerly winds. By September they have exploited fields of milkweed as far north as southern Canada and, in places, built up large populations. Then they begin to disappear from the fields; those that have not yet reached adulthood are killed by the cold, while many of the adults fly south, taking advantage of autumn winds from the north.

No one has actually tracked migrating milkweed bugs, although they have been seen to take off and fly upward into favorable winds until out of sight. It isn't possible to band small insects the way that birds can be banded, yielding much of what we know about bird migration (although marking migratory monarch butterflies has met with some success). Nor is it practical to follow migrating insects with planes (except in the case of huge swarms of locusts). So most of what we know about the migrations of milkweed bugs is based on carefully executed laboratory experiments.

Seeking to learn more about the stimuli that cause the bugs to undertake migratory flight in the spring and fall, but to forego it when conditions are uniformly summerlike, Hugh Dingle, then at the University of Iowa (now at the University of California at Davis), devised a method of "flight-testing" milkweed bugs grown under controlled laboratory conditions. A short stick was glued to the back of each bug, and the stick clamped to a stand so that the bug could be suspended over a black surface, with a 100-watt light bulb above. Flight-testing consisted of a series of stimuli of increasing intensity. First, they were lifted off the surface. If that didn't cause them to fly, they were tapped with another stick; then, successively if they didn't fly, a jet of air was applied, they were moved back and forth, or their wings were opened manually. Flight that persisted for thirty minutes or more was regarded as migratory, while short-duration, easily elicited flights were equated with movements that might occur in the field from one milkweed plant to another. As might be expected, newly emerged adults didn't fly readily, nor did very old individuals. At about eight to ten days of adult age, many individuals flew persistently, and Dingle estimated that they might, in nature, have covered at least sixty miles, probably much more with a tail wind.

Having learned this, it was possible to culture the bugs under different conditions of day length, temperature, and food availability, and then to flight-test the adults. In fact, all of these were found to influence the tendency to migrate. Young adults mate, and females, if nonmigratory, proceed to lay their eggs within a few days. But migratory females show suppressed development of their ovaries and do not lay eggs until after migration. When bugs are reared under a day length of sixteen hours and a night length of only eight hours (summer conditions), they show little tendency for prolonged flight and the females begin laying eggs soon after mating. But when days and nights are of equal length, egg-laying is much delayed and prolonged flight is much enhanced. Decreasing the temperature (within certain limits) also prolongs the period of egg-laying and thus the time available for flight. Tendency to migrate can also be increased (for a time) by starvation. Thus, under summer conditions of long day length, high temperatures, and plenty of food, the bugs tend to stay put. But in the autumn, cool temperatures, declining day length, and milkweed plants that have passed their prime, all conspire to delay egg-laying until after many of the bugs have taken flight and moved southward. In the spring, suboptimal food (lack of seeds) may be the principal stimulus producing the northward migration.

So there appears to be an inverse relationship between migration, on the one hand, and feeding and egg-laying on the other. Migration (when it occurs) involves relatively young individuals, the females having ovaries that are not fully developed (a statement that will apply to most migrating insects). Migrants are said to be in "reproductive diapause," diapause being a state of arrested development. Hibernating insects undergo total diapause; their activities are reduced to zero and they undergo physiological changes that enable them to withstand freezing temperatures. Migrants undergo an arrested development of their reproductive organs and their tendency to feed only; this is terminated after they have arrived in an area where their offspring have a better chance of surviving.

It didn't take long for physiologists to look into the mechanisms that control migration and reproductive diapause. It was known that the development of the female's ovaries was under the control of a hormone secreted by the corpora allata, small endocrine organs just behind the head. This hormone is also responsible for suppression of the development of adult features in immature insects, so it is called juvenile hormone (often abbreviated to JH). It is the "Peter Pan" hormone of insects, keeping them immature as long as it is being produced. When they molt to the adult stage after a genetically determined period of development, they do so in the absence of JH. But in the adult, the corpora allata become gradually reactivated, and the same hormone is responsible for development of the ovaries as well as various aspects of adult behavior. It is, in fact, the major sex hormone of insects. (Some of the earlier research on insect hormones was reviewed in Chapter 9 of my book *Life on a Little-Known Planet*).

If the corpora allata of an adult insect are removed surgically, there is no development of the ovaries. But if one implants active corpora allata from another female, ovarian development is restored. Removal of the corpora allata also suppresses flight activity, but this can be restored by applying JH to the insect. So it appears that JH is responsible for both prolonged flight and ovarian development, even though these activities are mutually exclusive. This posed a dilemma that has since been resolved by Mary Ann Rankin, of the University of Texas, and Lynn Riddiford, of the University of Washington. These researchers reared milkweed bugs under various conditions, flight-tested them by Dingle's method, then extracted blood samples and determined the level of juvenile hormone.

Under long-day and high-temperature conditions, JH concentration builds up rapidly and induces full development of the ovaries in about ten days of adult life. But at lower temperatures and shorter days, buildup of JH is much more drawn out, and within a few days reaches a level sufficient to stimulate long flights but insufficient to induce ovarian development. Eventually, hormone titer becomes high enough to cause the insect to cease flying and undergo rapid ovarian development—but by this time the insect is likely to be in a more favorable environment. Lack of food or poor quality food also results in lower JH concentration, thus inducing flight, though if starvation persists flight ceases. These statements probably apply to many insects, but the milkweed bug just happened to be there on the laboratory shelf, picking away at its seeds, and proved just the right animal to provide answers to critical questions.

The small milkweed bug, *Lygaeus kalmii,* despite its resemblance to the larger species, responds quite differently to both internal and external cues. Application of JH results in no increase in flight activities, and evidently flights are primarily short-range responses to food condition and availability. The corpora allata primarily control ovarian development, and egg-laying typically follows mating fairly promptly. On the other hand, decreased day length and cool temperatures produce a total diapause, or hibernation, such that these temperate-zone insects are able to occupy their range permanently.

These ideas are not without application to bird migration, as Hugh Dingle has pointed out. Some birds are well adapted for survival in cool regions, too, although of course they do not hibernate but flock to available food sources. Others, of more tropical groups, migrate, and do so at a nonreproducing stage. Again, it is hormones that are involved in the induction of migration, and these are produced in accordance with cues such as day length. To some extent birds and insects use the same kinds of orientation cues during migration—landmarks, light-compass, and perhaps geomagnetism—but by and large insects make much more use of air currents. Also, the much shorter lives of insects means that the same individuals don't usually make a return flight.

But it does appear that general theories of animal migration are now emerging. Many animals besides birds and insects migrate, of course, including many fish, whales, and grazing mammals. The lowly, "unimportant" milkweed bug is playing an important role in the development of these ideas. Recent years have seen no decline in the number of scientific articles published on the bug, and doubtless it still has a great deal to teach us.

13

The Tobacco Hornworm

I suspect that when most people dig into the recesses of their minds for their earliest childhood memories they come up with scenes of kittens, puppies, or hamsters. My earliest memories are of tobacco hornworms, and how delightfully they pop and ooze between bare toes. Picture a tobacco farm in the Connecticut Valley, with kids walking up and down the rows looking for big, green caterpillars and executing them by the most primitive of control measures. I was amused to note just the other day that one reasonably up-to-date book on insects still comments that "when labor is cheap and plentiful, hand-picking of these worms is fairly effective." We were certainly cheap, though we did not seem very plentiful when confronted with forty acres. In any case the hornworms were insignificant compared to the succession of hailstorms that eventually drove my father to other crops and finally to bankruptcy.

Tobacco hornworms also feed on tomatoes (along with a very similar species, the tomato hornworm), so they are well known to gardeners everywhere. Since they are green and have a pattern of pale lines that breaks up their outline, they are almost impossible to spot until, one day, one notices that the plants are mostly defoliated and huge "worms" cling to the shredded plants. There is another reason why one rarely sees them until they are virtually fully grown. When they are young they are very small indeed, and have small appetites. Specialists of the U.S. Department of Agriculture tell us that before their first

molt, hornworms eat an average of only 0.1 square inch of leaf;
between the first and second molt, 0.36 square inches; between
the second and third, 1.86; between the third and fourth, 16.32.
But after that, following the fourth molt and before the fifth molt
to the pupal stage, they consume an average of 201 square
inches of leaf—90 percent of their entire intake. So it is not surprising that one rarely notices them until they have pretty well
eaten about all they are going to eat.

Fully fed hornworms drop to the ground and form their
pupae in the soil. Unlike many caterpillars, they spin no cocoon
at all. Next summer (or later in the summer, in the South), great
gray hawk moths emerge, flying by night to feed, to mate, and to
lay their eggs on tobacco or tomato plants. The moths are
equipped with tongues coiled like watch springs, which when
uncoiled are longer than their bodies, able to penetrate the deepest of blossoms for nectar. The moths have six pairs of yellow
spots on their abdomens, hence the scientific name *Manduca
sexta*—*sexta* meaning "six-fold," *Manduca*, the Latin word for
"glutton." No one who has watched the caterpillars fatten on his
tobacco or tomato plants will question the appropriateness of
the latter!

For most persons, hornworms do not evoke nostalgia, as
they do with me, but rather feelings of anger and revulsion.
There is, however, much to be said for them. Both tobacco and
tomato foliage is loaded with powerful alkaloids: nicotine and
tomatine, respectively. Nicotine has, of course, been used as an
insecticide for many years, and still is. Not many insects or herbivorous mammals will touch these plants. Yet hornworms are
not only able to tolerate these poisons, but they use them as
feeding cues—in nature they will feed on nothing else (except for
a few closely related plants with similar alkaloids). Tobacco and
tomato have presumably evolved these poisons as a means of
deterring feeding by herbivores. Having overcome the plants'
defenses, hornworms have assured themselves of a rich source
of food where there will be little competition from other species.

How do tobacco hornworms manage to ingest so much nicotine without being poisoned? Many insects that feed on plants
containing poisons have evolved means of detoxifying these
substances; that is, they are able to convert them in their bodies

to nontoxic byproducts. This doesn't seem to be true of tobacco hornworms. Researchers at the North Carolina Agricultural Experiment Station have shown that the nicotine passes through the digestive tract unchanged and is promptly excreted. When larvae are fed radioactive nicotine, almost all of it can be detected in the feces within four hours. Although a little is absorbed in the blood, it is never enough to poison the hornworms. Even when nicotine is injected into them, it is eliminated within a few hours after causing a merely temporary paralysis.

Caterpillars taste primarily with minute sense organs on their maxillae (paired appendages just behind the mandibles). Researchers have found that if the maxillae are removed surgically, hornworms can be induced to feed and develop more or less normally on dandelions, burdock, and other plants they would never touch in nature. So the selection of tobacco and tomato seems to have little to do with nutrition, but rather with an evolutionarily "acquired taste" for their characteristic alkaloids. Surprisingly, tobacco hornworms can be reared successfully on artificial diets containing no plant material at all, diets consisting of wheat germ, casein, brewer's yeast, sugar, agar, and a variety of lesser ingredients. When reared on this medium, they can be transferred (when partly grown) to cabbages, plantains, and other abnormal food plants, where they feed readily even with intact maxillae. Apparently their taste receptors undergo a decline in sensitivity, so that they will accept many different plants.

The fact that hornworms can be easily reared in the laboratory on an artificial diet renders them especially useful to teachers and to research workers. It is possible to purchase prepared hornworm diet from a variety of sources, and some biological supply houses sell "hornworm kits," with everything necessary for students to keep "pet" hornworms in bottles and watch them grow and transform to pupae and to adult moths. With appropriate ligations, the application of synthetic hormones, and the like, it is possible to do some fascinating experiments with these homely creatures. They are even large enough that surgery on their nervous system or other internal organs is not all that difficult.

Oddly, hornworms reared on an artificial diet are not green

but blue. A simple explanation might be that they are not green simply because they obtain no green pigment from leaves. But why blue? In fact hornworms are intrinsically blue as a result of a blue bile pigment in their system. From the plants they obtain a yellow pigment that complements the blue to make them green. But in the absence of plant tissue they reveal themselves to be "blue bloods" more truly than you, I, or the Cabots or the Lowells.

Hornworms are also sometimes black in color. Tomato hornworms that develop under cool conditions turn out to be black, and although this is not true of tobacco hornworms, a black mutant strain of tobacco hornworms has appeared spontaneously under laboratory conditions. Black tobacco hornworms can also be produced in the laboratory by ligating the necks of normally green worms at appropriate times, or by surgically removing a pair of endocrine glands, the corpora allata. In either case one shuts off the secretion of these glands, the juvenile hormone (JH). While the major function of this hormone is to suppress the development of adult features until larval feeding has been completed, it also plays a role in controlling pigmentation of the skin. If the dark coloration is indeed a result of reduced juvenile hormone, then it ought to be possible to perform the opposite experiment and apply juvenile hormone to young caterpillars of the black strain and cause them to be green. Sure enough, it works.

So evidently the black mutant has a minor malfunctioning in the production of juvenile hormone at certain periods, not enough to affect metamorphosis but enough to influence pigmentation. Many caterpillars as well as many grasshoppers have the capacity to be either light or dark in coloration, and in most cases it can be demonstrated that the amount of JH at critical periods is responsible. Often some environmental cue influences the endocrine system to cause the deposition of dark pigment in the skin: cool weather, day length, or crowding, for example. In some cases, having two color forms permits insects to blend in with local background colors. This hardly seems to be the case with hornworms—black caterpillars in green foliage must be fairly conspicuous to predators. But hornworms have provided excellent experimental animals for study of the physiological basis of different color forms.

Explaining the precise chemical basis of these changes is, of course, a more difficult matter, a matter being explored especially in the laboratories of Lynn Riddiford and James Truman, a wife-husband team at the University of Washington in Seattle. But that is only one facet of the research on hornworms being done in their laboratories. It is well known that fully grown, fully fed caterpillars undergo a decline in juvenile hormone in their blood. They are, however, under the influence of a second hormone, the molting hormone, which in turn is under control of a master hormone in the brain. When tobacco hornworms reach a weight of seven grams, they enter a "wandering" stage, which leads them to a suitable place to molt to the pupal stage, an event occuring four days later. Both of these events are triggered by the production of brain hormone, which is under the influence of a "biological clock" that is "set to go off" in the early hours of night and then again four days later. Larvae that are fully fed cannot "wander" until the "clock" permits the release of brain hormone; if they miss the appropriate timing one day, they have to wait until the next day. All of this can be demonstrated, and modified, by ligating the head and thus shutting off the flow of brain hormone at specific times, one of many recent demonstrations of the influence of hormones on insect behavior.

Ligation is a simple matter, done by tying off the neck tightly with a piece of dental floss. (It is not true that sophisticated experiments always require elaborate instrumentation, but they do require a lot of know-how, in this case as to when to make the ligation and how to interpret the results.) Many other things can be done with hornworms. For example, if the neck is ligated, one can still produce a molt by injecting into the body brain substance from another individual that is producing brain hormone. By cutting hornworm brains into small pieces, Lynn Riddiford and her associates have been able to determine in what part of the brain the hormone-secreting cells are located. It is also possible to culture pieces of hornworm epidermis *in vitro*, that is, in special media apart from the animal. Then, by introducing hormones in various relative concentrations, one can observe how they influence the metamorphosis of the epidermis. Synthetic juvenile and molting hormones are now available from several sources, rendering many experiments possible that were unthought-of only a few years ago. All of this is far more than just "tinkering with bugs"; it is getting at the sources of insect development and behavior, with all that that implies in terms of making insects more subject to our manipulations, either toward their demise or increase.

One of the many neat things that have been done in Riddiford and Truman's laboratory is to make use of the effect of juvenile hormone on coloration of tobacco hornworms to develop a bioassay for the hormone, that is, a way of measuring the level of hormone in a blood sample to be tested. Briefly, appropriately ligated hornworms are anaesthetized and the sample to be tested is applied to a standard area of the abdomen. Normally a ligated larva would be black, but when juvenile hormone is applied, a patch of integument turns green, and the size of the green spot, over a two-day period, is indicative of the concentration of JH in the sample. The precise relation of spot size to hormone concentration was first worked out by applying known dosages. Then the concentration in samples of unknowns could be determined by reference to this scale. It was by use of this bioassay that the relation of hormone level to migration and to ovarian development was worked out in the milkweed bug, as discussed in the previous chapter. Thus two "laboratory insects," though wholly unrelated, became mutually useful.

As tomato and tobacco growers are well aware, hornworms are fairly often found to be covered with white, elliptical, silken objects. These are the cocoons of a small parasitoid wasp, *Apanteles congregatus*. Infected hornworms should not be destroyed; they will die anyway, and the cocoons will produce a new crop of parasitoids. These wasps were first discovered by pioneer entomologist Thomas Say, who will be a leading character in Chapter 15. In June 1822, Say "obtained eighty-four individuals of this species from [a hornworm]." He appropriately called the species *congregatus*, for in fact it normally occurs in

---

All too familiar to farmers, the tobacco hornworm is nevertheless the "pet" of many researchers.

P. EADES

"congregations" of a few to a hundred or more individuals in a single hornworm.

The wasps (not much more than an eighth of an inch long) dart and hover about the plants and now and then jump on a hornworm, usually one that has just hatched or is only a few days old. The hornworm defends itself by thrashing its body about, but it takes the wasp only a few seconds to pierce the body wall and lay several eggs inside. The parasitoid larvae live inside the hornworm for about a week, living on nonvital tissues, then puncture the body wall and emerge to spin their cocoons on the top or sides of their host. How many emerge from a single hornworm depends on how many batches of eggs it has received. Up to the time the wasp larvae emerge and spin their cocoons, the hornworms feed slowly and molt once or twice, but after that they remain quiescent, feed no further, and eventually die without completing their development.

Apanteles congregatus is an example of a native insect that attacks native insects—it has never been used in biological control as such, but is one of a great horde of unsung insects that play a role in keeping populations of pest species at reasonable levels. It does equally well on both tobacco and tomato hornworms. But oddly, when tobacco hornworms are feeding on certain kinds of tobacco, most notably the "dark-fired" variety that is grown extensively in Kentucky and Tennessee, *Apanteles* doesn't do well at all. Apparently the level of nicotine is higher than the wasps can take, although the hornworms do well enough.

Lynn Riddiford and her associates have shown that the time the *Apanteles* emerge from their hornworm hosts is determined by the number of parasitoid larvae developing in that host. When very young hornworms are parasitized by only a few *Apanteles*, the latter generally complete their development and emerge when the host has molted only twice; but if there are a hundred or so, the host may have molted four times before the parasitoids emerge. In no case will the caterpillar molt to the pupal stage, as it normally would at the fifth molt. If the load of parasitoids is *very* high, say two or three hundred, they may in fact emerge after a fifth (or rarely even a sixth) molt—but the hornworm remains a caterpillar with "extra" stages rather than becoming a pupa. This is obviously adaptive for the *Apanteles* larvae, as the more there are, the more competition there is for the limited food; so it is to their advantage to develop more slowly and to permit the host to continue to live and to feed.

But how is it brought about? In fact, hornworms are prevented from molting to the pupal stage because, through some mechanism not fully understood, the parasitoid larvae induce a very high level of juvenile hormone in their hosts. Bursts of molting hormone occur at intervals, but the level of juvenile hormone remains such that juvenile features continue to be imposed, and a pupal molt does not occur. In fact, the *Apanteles* larvae depend upon these bursts of molting hormone to complete their own development and emergence from the host. When Riddiford and her associates applied juvenile hormone to parasitized hornworms, emergence of the parasitoids was delayed or even prevented, since the delicate hormone balance required for their emergence had been disturbed.

Synthetic substances with hormonal effects have been suggested and to a limited extent used in insect control. Harvard biologist Carroll Williams calls these "third generation pesticides," the first generation being the "old fashioned" inorganic pesticides such as arsenic, the second the complex organic substances so popular today. One "state of the art" pesticide being recommended for control of caterpillars of many kinds is ETB, an anti-JH agent that suppresses juvenile hormone and thus induces precocious metamorphosis. Studies in Riddiford's laboratory have shown that ETB is relatively ineffective against parasitized hornworms, apparently because the parasitoid larvae maintain an abnormally high level of JH and thus "protect" the hornworm from this agent. Obviously, the use of hormones and antihormones to control insects is not a simple matter, considering the complex ways that parasitoids alter the physiology of their hosts to fulfill their own requirements.

When Nancy Beckage, working with Lynn Riddiford, tried rearing a different species of *Apanteles* on tobacco hornworms, she found that these parasitoids could not always suppress molting by the hornworms after the wasps emerged. Many molted to form larval-pupal intermediates, indicating that the parasitoids had only incomplete control over the physiology of their hosts. Also, many fewer parasitoids developed successfully, as compared to *congregatus,* which is obviously much better adapted to tobacco hornworms. What species a parasitoid is able to attack successfully obviously depends in some cases on its ability to regulate and exploit the hormone levels of its host as well as its ability to tolerate whatever plant-derived substances are circulating the blood of the host. As biological control of insects becomes a more exact science, these are matters that will have to be explored more thoroughly.

The research that has been done on hornworm caterpillars I have reviewed in only the most superficial way. Equally extensive research has been done on the adult hawk moths by Bernd Heinrich, formerly at the University of California at Berkeley but now at the University of Vermont. These are drab but impressive moths, with stout, tapered bodies, long, angular fore wings and small hind wings. Flying at night, they hover about flowers and probe them for nectar; in fact they spend most of their active hours in flight, finding their mates there and seeking plants on which to lay their eggs. By inserting thermocouples into the flight muscles of free-flying tobacco hornworm moths, Heinrich showed that they maintain a temperature of 38 to 42 degrees Celsius (slightly over 100 degrees Fahrenheit). Obviously, these are not "cold-blooded" animals as they are reputed to be.

If such a temperature is required for flight, how do they attain such a temperature in the cool of evening? In fact they "warm their motors" by rapid contractions of flight muscles, causing the muscles to "shiver." Within a few minutes enough heat is produced to permit flight. The dense covering of scales provides insulation, so that much of the heat produced is retained in the thorax. If the scales are removed, the moths can still warm up, but it takes longer.

Since the flight muscles produce so much heat, how do the moths avoid overheating during prolonged flight? Heinrich showed that they accomplish this by increasing the flow of blood through the thorax and into the abdomen, which is larger than the thorax and more thinly scaled. Insects are unable to cool themselves by perspiring, as we do, but simple convection from the abdomen during flight causes the blood to cool before it returns to the thorax through the dorsal vessel, the so-called insect heart. Thermocouples inserted into the abdomen of hawk moths during flight show that the temperature is appreciably lower than in the thorax and only slightly above air temperature. If the dorsal vessel is ligated so that blood cannot flow back into the thorax, the flight mechanism overheats and the moths cease flying. When the moths fly under cool conditions, the rate of blood flow can be decreased, so that most of the heat remains in the heavily insulated thorax. Thus the moths are able to avail themselves of a wide range of air temperatures even though their flight muscles work efficiently over a very narrow temperature range.

Once again the tobacco hornworm, in this case the adult moth, provided the ideal experimental animal: large, easy to obtain and to rear, and providing a paragon of fundamental functions. It is now clear that many insects, especially larger ones, require specific, rather high temperatures for flight, but they are able to "warm up" in a manner similar to that of the hawk moths. Fueling themselves with nectar, they are able to propel themselves swiftly even in cool air temperatures. Bumble bees (also among Bernd Heinrich's special pets) can even fly when the air temperature is barely above freezing, an adaptation that permits them to fly high in the tundra of mountaintops or of the far North.

Honey bees also fly in cool temperatures and are, in general, more successful than bumble bees in humid tropical or in desert environments. Honey bees are able to fly at remarkably high temperatures—up to 46 degrees Celsius (115 degrees Fahrenheit). They accomplish this by using an additional method of cooling: they periodically regurgitate some of the nectar from their crop out their tongues, resulting in evaporative cooling of the head and thorax by several degrees. Bernd Heinrich calls this "keeping a cool head." Because of their ability to thrive in so many habitats, honey bees are supreme competitors with other bees and tend to replace native bees in many parts of the world. Hawk moths have also been seen with droplets of nectar on their mouthparts, and it is quite possible that they, too, may use evaporative cooling if exposed to high temperatures.

The first person to demonstrate that flying insects are far from "cold-blooded" was British entomologist George Newport, who in 1837 concluded that "every individual insect when in a state of activity maintains a separate temperature of body considerably above that of the surrounding atmosphere, or medium in which it is living. . . . " This is the same George Newport we met as a pioneer student of the life cycles of blister beetles, in Chapter 8. Newport used mercury thermometers, and his crude measurements were not always on the mark. Nowadays researchers use thermocouples, electrodes, potentiometers, and other sophisticated equipment, and the subject has proved a most fertile one, recently summarized in a volume edited by Bernd Heinrich, *Insect Thermoregulation.*

Thus the tobacco hornworm has become something of a "key species" for physiological investigations, and what it has taught us can be applied to many other insect species. A "pest" it may be, but an admirable and useful one—chiefly for specialists, to be sure, but it does add excitement to the lives of tomato growers; and whether one objects to its penchant for tobacco depends upon whether one is a smoker or has stock in R. J. Reynolds. For myself, I rather look forward to them each year in my tomatoes; they are, in a way, heroic creatures. And if they go a little too far in their gormandizing, well, I confess I still enjoy squashing them with bare feet, as I did as a child.

14

Enjoying Insects in the Home Garden

The growing of vegetables in one's backyard has much to recommend it. Some thirty-eight million Americans have such gardens, altogether occupying nearly two million acres. The flavor and food value of freshly picked vegetables so far exceed those of the store-bought equivalent that gardeners look forward eagerly to each year's harvest— and spend cold winter days planning next year's. (I always order my seeds from the catalog on the day the IRS forms arrive; it takes away some of the pain.)

It is claimed that despite the high cost of food these days, one doesn't really save money by having a home garden—if one counts the cost of one's own labor. But who wants to count that when gardening is such good exercise and so full of challenges and rewards? What will the summer's weather be like? Drought? Hail? Early frost? Will we have enough sweet corn to invite friends for a corn roast? How many canning jars will be needed?

Then, of course, there are insects to think about. Will flea beetles be abundant this year? Will leaf miners decimate the beet greens? Will the cabbage butterflies find the broccoli? (Of course, they always do.) Shall I ring the garden with marigolds to keep the grasshoppers out? (Doesn't work; they love marigolds.) Shall I sacrifice my ideals and stock up on insecticides? Or shall I plant a little more than we need and simply enjoy the insects, the rabbits, the birds?

There is much to be said for the last suggestion. Our garden is not far outside our picture windows, and as we drink our coffee we can watch the robins and grackles slipping into our strawberry patch and emerging with their beaks smeared with red. Now and then a squirrel stares into the patch, then dives in, emerges with a berry, and takes off for the nearest tree. Like the rabbits that nibble the peas, he has learned that we may pursue him. It is fun for all; of course we never catch anything.

Each year I look forward to the insect inhabitants of the garden. I don't begrudge them their share; they amuse us, inform us, and often stimulate our sense of beauty. If they take what I consider more than their share (a fairly rare event) I don't mind being a bit brutal. No animal should multiply so as to destroy its environment (though, being a member of the species *Homo sapiens*, I may have no right to say that).

As I write this I am rearing some zebra caterpillars that were skeletonizing the leaves of our roses. (We do make space for flowers, which after all are food for the spirit.) The zebra caterpillar is very beautiful indeed: its head is orange, its body yellow, with three pairs of black, longitudinal stripes, each pair separated by a white streak. In color it rivals the orioles that are nesting in our cottonwoods. One does, of course, need a microscope to fully appreciate the beauty of zebra caterpillars, but that is an essential item in the household of everyone who admires insects. The caterpillars don't remind me much of zebras. I prefer the scientific name, *Ceramica picta*, which I would translate from the Greek and Latin to mean "a painted earthen vessel."

It has been a good year for asparagus beetles. I am glad; we had a good harvest, and the beetles are most decorative on the bushy flowering stalks. There are two kinds, and we have them both. One, officially called *the* asparagus beetle, has a black head, antennae, and legs, with an orange collar and blue-black wing covers bordered with orange and bearing six light yellow spots—as elaborate a color pattern as one could design. The other, called the spotted asparagus beetle, is wholly orange except for black "polka dots" on its wing covers. The two kinds have similar defensive behavior: if the bush is shaken they simply release their grip and drop to the ground, a response called "thanatosis" in the scientific literature, where suggestive terms such as "playing dead" are frowned upon. If captured, both emit high-pitched squeaks, presumably addressed to birds and other predators, which may drop them in surprise.

Neither species feeds on anything other than asparagus, and hardly anything else in the insect world will eat asparagus. Is there some chemical in the plant that repels herbivores? How does it happen that two related species of beetles can live together on the same plant at the same time—a seeming contradiction to the rule that complete competitors cannot coexist? What is the significance of the fact that the eggs of the asparagus beetle are laid in rows in an erect position, while those of the spotted asparagus beetle are laid singly, flat against the stems? How did two related species happen to evolve such brilliant and such different color patterns?

It appears that the larvae of the asparagus beetle feed freely on the foliage, while those of its spotted cousin live primarily within the seeds. Thus to a certain extent they share the plants and are not complete competitors. Perhaps, with some research, I could answer the other questions. But these are difficult questions, without simple answers. And there are so many other unanswered problems within a few yards of our back door!

To consider only the matter of food specialization: we have plenty of "finicky eaters," gourmets, if you like: asparagus beetles, Mexican bean beetles, tomato hornworms, and others. We also have plenty of insects that will eat almost anything: cutworms and grasshoppers, especially. It is intriguing to speculate on the advantages of being a specialist, when the disadvantages are so obvious. Clearly, if I don't plant beans, I won't have bean beetles, and if nobody plants them (and there are no wild ones) the bean beetles will become extinct. Specialist feeders are closely linked to their hosts and must suffer when the host declines, just as the chestnut weevil disappeared when the chestnuts vanished from eastern forests as a result of the blight.

Yet there are clear advantages in being a specialist, for specialists have of necessity evolved responses to specific cues of their host plants (usually chemical) and use these cues to find their favored plants readily among the myriads of plants in nature and in gardens. Often the plants they specialize on have repellent or even poisonous chemicals. Volatile mustard oils in members of the cabbage family, for example, repel many insects, but enable cabbage butterflies to zero in on plants on which their larvae thrive. An alkaloid called tomatine, in tomatoes, is toxic to many leaf-feeders, but hornworms have evolved digestive systems that get rid of the toxins without suffering any ill effects. By having overcome certain repellent or toxic substances, host-specific insects have the plants pretty much to themselves, without serious competition. Quite often they make use of the toxins to render themselves distasteful to predators. This is what the milkweed bug and the monarch butterfly have done with the cardiac glycosides of milkweeds. Perhaps that is the explanation of the predominantly orange coloration of asparagus beetles, orange being the "warning color" of many distasteful insects.

In contrast to specialist feeders, generalists can find food almost anywhere and will often eat weeds and wild plants as well as garden plants. But they tend to avoid or do poorly on plants containing toxins or repellents, since they haven't developed the ability to overcome the specific chemicals of defended plants—and it would hardly be possible to evolve mechanisms for overcoming such a diversity of chemicals. Even in bad grasshopper years, tomatoes remain inviolate.

Of course there is no such thing as a "complete generalist"; even gypsy moths don't eat everything in sight, quite. And there are few that confine their feeding to a single species of plant, although there are many that feed on a few related plants, usually members of the same family. Cabbageworms (the larvae of white cabbage butterflies), for example, feed readily on cabbage, broccoli, brussel sprouts, and cauliflower—all members of the cabbage family having similar leaf chemistry. Parsleyworms (the larvae of black swallowtail butterflies) feed on carrot tops, parsley, dill, and parsnips, all members of the carrot family (Umbelliferae).

Parsleyworms are most welcome denizens of our garden. They are elegant creatures, transversely banded with green and black and ringed with orange spots. When disturbed, they erect a pair of orange "horns," actually eversible glands that secrete a

The Colorado potato beetle holds an important place in the history of entomology.

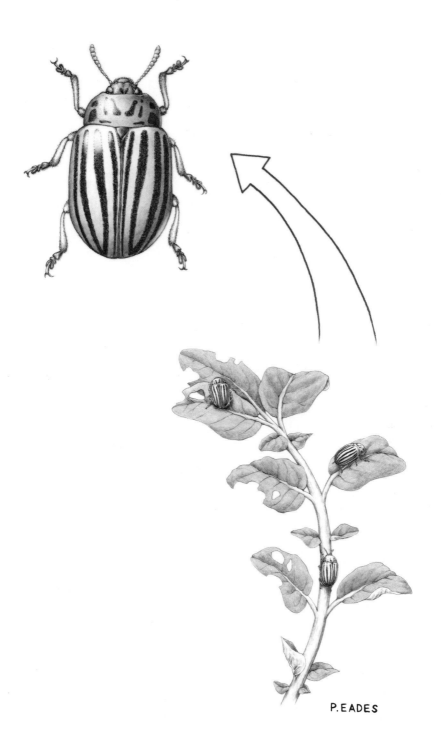

P. EADES

repellent substance that smells like rancid butter—in fact, it is butyric acid, the essence of rancid butter. It is an enjoyable experience to collect the mature larvae or chrysalids and rear the butterflies indoors, where one can follow the expansion of their glossy, spangled wings. We always release them outdoors, of course, where they can find mates and plants on which to lay their eggs. Their larvae are seldom abundant; it would take a lot of them, all season, to do as much damage to our carrots as the local rabbits can do in one evening.

Parsleyworms are fastidious feeders, requiring food containing certain essential oils, such as methyl chavicol, which is found principally in members of the parsley family. They will even attempt to eat paper if it is soaked in these oils. On members of the parsley family, they can feed without serious competition from other insects, since parsleyworms have evolved the ability to thrive in the presence of psoralens, substances that deter feeding by most other insects by binding DNA in the presence of ultraviolet light. G. Wayne Ivie and his colleagues in the U. S. Department of Agriculture have recently fed tissues treated with carbon-14-labeled psoralens to fall armyworms (which are very general feeders) and to parsleyworms. They found that parsleyworms rapidly detoxify the poisons in their midgut, so that they do not enter the body fluids to any great extent; within 1.5 hours, 50 percent of the carbon-14 passes out with the feces. By contrast, fall armyworms accumulate more psoralens in the body tissues, and within 1.5 hours only 1 percent of the carbon-14 has appeared in the feces. So it is not surprising that, despite the appetite of armyworms for plants of many kinds, they do not flourish on parsley and related plants.

It is interesting that some members of the parsley and carrot family are relatively unpalatable to parsleyworms. Paul Feeny and his colleagues at Cornell University have shown that cow parsnip and angelica have evolved certain modifications of the form of the psoralen molecule that cause a reduction of the growth rate and fecundity of parsleyworms. They cite this as an example of coevolution: parsleyworms first evolved a means of overcoming the effects of psoralens, in members of the carrot family, and later certain members of this family evolved a modification of the molecule that parsleyworms could not handle.

Will parsleyworms further evolve the ability to overcome this novel plant defense? In a few hundred years, perhaps a few thousand, the answer should be apparent.

On the other hand, some plants of the parsley family that grow as wildflowers in woodlands lack psoralens, which are ineffective as deterents in the absence of plenty of light. Unlike most umbellifers, these plants are attacked by a variety of generalist feeders.

Parsleyworms, cabbageworms, and similar insects were the subject of a 1964 article that has become something of a classic of entomology: "Butterflies and Plants: a Study of Coevolution." The authors were Paul Ehrlich, of Stanford University, and Peter Raven, of the Missouri Botanical Garden. The science of plant-insect relationships has since blossomed into a major field, the subject of several books and innumerable scientific articles. Once these relationships are better understood, it may be possible to breed varieties of crop plants that either lack the chemical cues required by pest species or that have repellent or toxic properties with respect to these insects. To some extent this is already being done. For example, plant breeders have for some years been developing kinds of wheat resistant to the notorious Hessian fly, a stem-infesting insect reputed to have been brought to this country in straw bedding used by Hessian mercenaries employed during the Revolutionary War. Resistant varieties provide a theoretically perfect insect control, requiring no intervention with insecticides. Unfortunately, insects can sometimes overcome, in a few generations of evolution, resistant factors that have been bred into these stocks. This has happened with the Hessian fly, requiring a continuing program of plant breeding.

That, incidentally, is one of the strongest reasons for preserving as much of the wild plant world as possible: the genetic material needed for the breeding of resistant stocks may lie in wild relatives of wheat, corn, and other crops. The rapid extinction of species and locally adopted populations that is occurring as a result of widespread habitat destruction is not pleasant to contemplate as we look forward to feeding the crowded world of the future.

Many of the seeds available to home gardeners are, in fact,

those of varieties that have been developed for resistance to various diseases and insects. However, neither our basic knowledge of this complex subject nor our technology has advanced to the point that we need fear that next year we will miss the cabbageworms, the parsleyworms, the asparagus beetles, and all the other insects that provide half the joy of gardening.

The growing of potatoes in the home garden is not especially practicable, but I often plant a few simply because I like to have a few Colorado potato beetles around. There are few insects that have played so prominent a role in the history of entomology. The Colorado potato beetle (often called simply "the potato bug") is our gift to Europe—a recompense, so to speak, for their gift of the asparagus beetle, the Hessian fly, the cabbageworm, the gypsy moth, and so many others. According to Reece Sailer, of the University of Florida, more than 1,300 "foreign" insects have become established in the forty-eight contiguous states, some 800 of these deserving to be called "pests." But the Colorado potato beetle is a native of Mexico and the southwestern United States that has spread over most of North America and much of Europe, producing a wave of devastation of one of humankind's most basic crops, the potato.

Pioneer entomologist Thomas Say first discovered the potato beetle in 1824, calling it *Doryphora 10-lineata* (literally, spearbearer with ten stripes) (it is now called *Leptinotarsa decemlineata*). Say found it "on the Upper Missouri," and although he did not record the host plant, we now know that the beetles originally fed on a common western weed known as buffalo bur (*Solanum rostratum*). It was not until the early settlers brought the potato to the West that this once obscure insect ran rampant. Potato (*Solanum tuberosum*) is a close relative of buffalo bur, with similar leaf chemistry, and the beetles found it thoroughly palatable. They spread from one potato patch to another, often destroying the whole crop. By 1964 they had reached Illinois, where state entomologist Benjamin Dann Walsh remarked that they marched through the state "in many separate columns, just as Sherman marched to the sea." By 1874 they had reached the Atlantic coast—a remarkably rapid spread for a relatively weak-flying insect, suggesting that some had "hopped a train east."

But we should not pass up Walsh so briefly. He was a colorful person, who went about the fields in a long cloak and a tall, cork-lined hat. When he caught an interesting specimen he impaled it on a pin and attached it to the cork lining of his hat. Late in life, while walking down a railroad track reading his mail, he was overtaken by a train and injured in such a way that he had to have a foot amputated. He looked forward to having it replaced with a cork foot, he told his wife, so that "when I am hunting bugs in the woods I can make an excellent pincushion out of it."

Walsh had been a classmate of Charles Darwin, and like Darwin had collected insects as a youth. But after he moved to Illinois, in 1838, he became a farmer and lumber merchant. When he was fifty he retired to devote himself to entomology. Although he lived only another dozen years, he published 385 articles on insects and coedited, with Missouri entomologist C. V. Riley, a short-lived, semipopular journal, *American Entomologist.* He was appointed Illinois' first state entomologist, and in this capacity made a careful study of what he called "the new potato bug."

Walsh had little use for those who tried to sell farmers quick remedies for every pest. The Union Fertilizer Company (whose secretary was appropriately named A. S. Quackenbosh) advertised a substance that it claimed was "sure death and extermination to the Cankerworm, the Curculio, the Apple Moth, the Potato Bug . . . and all descriptions of insect. . . ." Walsh replied: "The trouble with all such panaceas . . . is that we hear nothing of the ninety and nine cases where the [remedy] was applied and found to do no good. . . . Nothing is more certain than that there is no Royal Road to destruction of the Bad Bugs; and the only way in which we can fight them satisfactorily, is by carefully studying out the habits of each species. . . ."

Walsh followed his own advice, studying the biology of a number of important insects. But he by no means confined himself to "pest" species and their "control"; he made pioneering contributions to systematics and to ecology and evolution. He wrote of the restriction of insects to certain host plants and the possibility that species might evolve after continued breeding on a particular host. This was in 1868, not long after the publication of *The Origin of Species* and decades before plant-insect relationships became a major theme in entomology. A fuller account of the life of this remarkable man may be found in Arnold Mallis's 1971 book, *American Entomologists*.

But to return to the Colorado potato beetle, after still another diversion. So serious a threat was it during the last few decades of the nineteenth century that there was a desperate search for effective controls. Since (like asparagus beetles) these insects "play dead" when disturbed, one recommendation was to jar the plants so that the beetles would fall into a pan of kerosene or onto the ground, where, on a hot day, they might perish from the heat. Chemical insecticides were in a primitive state in the 1860s, and the first widespread use of chemicals to control insects is associated with the need to contain this pest. The arsenic-containing pigments Paris green and London purple were used at first, but later replaced by lead or calcium arsenate. Arsenicals came to be used for many insects, although nowadays they have for the most part been replaced by other substances. Incidentally, the first wide-scale test of the effectiveness of DDT, in 1941 in Switzerland, was against the Colorado potato beetle. Today one hears less about the beetle, as its natural enemies have to some extent caught up with it, and potato growers have established procedures for ridding themselves of this pest and several other insects and diseases.

It was in 1921 that the beetles became established in France. By 1935 they had reached Germany, and by the 1970s they were raising havoc as far east as Russia and Turkey. One of mankind's most staple foods was in jeopardy, and research proliferated in Europe as it had in the United States. More is now known about the feeding behavior of these beetles than that of perhaps any other insect.

One of the first persons to study the matter was Professor C. T. Brues of Harvard University, whose 1946 book *Insect Dietary* was an early effort to summarize knowledge of this intriguing subject. Brues kept twenty-four species of the potato family (Solanaceae) in his greenhouse and found that some provided acceptable food for Colorado potato beetles and some did not. Oddly, the adult beetles seemed attracted to tomatoes (which are also Solanaceae), although their larvae did not thrive on them. Brues called this "an extreme case of mistaken instinct." It now seems not surprising, since we know that tomato leaves contain a toxic alkaloid, tomatine. The native home of the tomato was quite different from that of buffalo bur, the original host of the potato beetle. Consequently, the beetles never evolved, as part of their genetically programmed behavior, the ability to discriminate between the leaf odors of their true host and those of a related plant that was toxic to them. Colorado potato beetles are also attracted to ornamental petunias, which like tomatoes are native to South America, but these solanaceous plants are also unsuitable as food for the larvae. When larvae are forced to feed on petunias, they vomit and become more or less paralyzed.

Ting H. Hsiao, of Utah State University, and Gottfried Fraenkel, of the University of Illinois, showed in 1968 that adult Colorado potato beetles, when given a choice, show a slight preference for laying their eggs on buffalo bur, as compared to potato, and a strong preference for laying their eggs on deadly nightshade (which belongs to the same genus as potato, *Solanum*). But deadly nightshade does not support larval growth at all! Hsiao and Fraenkel found that for egg-laying females, the physical nature of the leaves does not matter; they will lay eggs on hairy or spiny leaves, or even on glassware or paper towels, provided they receive the proper volatile chemical stimulation.

The problem of what these chemical stimuli are has been elusive. The leaves of most plants produce volatile "green odors," chiefly alcohols and aldehydes that have six carbon atoms in their molecules. J. H. Visser and his colleagues at the Agricultural University in Wageningen, the Netherlands, have shown that the antennae of Colorado potato beetles contain a variety of minute sense organs which are so designed that together they respond to a particular "medley" of these odors, specifically those of members of Solanaceae. Using a wind tunnel, Visser found that the beetles tend to fly upwind, particularly if the wind is enriched with the characteristic odors of solanaceous plants. Other kinds of plants produce other medleys of these same "green odors," as well as sometimes more specific odorous substances, and these similarly attract other insects. The green world is filled with subtle odors—but in fact a flying insect is "tuned in" to only certain complexes of molecules in certain concentrations; the rest it simply doesn't detect.

Finding the "right" plant is only the first step. Once on the plant, the insects (and/or the larvae that emerge from their eggs) will start feeding only in the presence of other chemical cues, such as sugars, cholesterol, and certain amino acids. However, they will not feed (or will develop slowly or even die) if certain repellent or toxic substances are present (such as tomatine or nicotine): unless, of course, they are specialist feeders able to tolerate or detoxify these substances (as, in this case, hornworms are, but not Colorado potato beetles).

Hsiao and Fraenkel, in the laboratory, tested Colorado potato beetle larvae on 104 species of plants of 39 families. They found that 80 percent of them were rejected by the larvae, presumably because they contain feeding inhibitors or toxins. But surprisingly, the larvae accept and grow well on certain nonsolanaceous plants such as milkweed and lettuce. However, these plants lack the necessary attractants for the adult beetles, so they are not accepted in nature. Obviously, host-finding and feeding are controlled by quite different factors—probably true of many host-specific insects. This is a "fail-safe" way of insuring that all goes well. But of course a modern garden contains plants that had their origins in many different parts of the world. Here they are crowded together and surrounded by a variety of weeds of even more diverse origins. Rather than accusing insects of occasional "mistaken instincts," one can but admire their ability to thrive in the midst of such a diverse and alien environment.

So there is something to be said for the Colorado potato beetle as a paragon of successful gormandizing. The prison-striped adults and the sluggish, corpulent larvae have found a place in the history of entomology and in the development of research both on ways of killing insects and on ways of understanding plant-insect relationships. Despite all this attention, the beetles are still there, waiting to infest our potatoes, joining the cabbageworms, the parsleyworms, the hornworms, the cutworms, the bean beetles, the asparagus beetles, and all the others—so that everyone may enjoy a miniature zoo in his backyard.

For gardeners everywhere, the late Cynthia Westcott wrote *The Gardener's Bug Book*, an encyclopedic treatment of all the "bugs" one might possibly encounter. Although ostensibly addressed even to "the organic gardener who shuns all chemicals and [to] the wildlife enthusiast who is worried about them," two early chapters are devoted to chemicals and how to use them. After listing alphabetically many hundreds of pests ("and a few friends"), she provides an alphabetical list of host plants with the important pests of each. It is an intriguing list, with all kinds of insects I could never hope to lure into my backyard, surely not the rednecked peanutworm or the longtailed banyan mealy-bug. But I am disappointed never to have attracted the harlequin cabbage bug or the imbricated snout beetle, both of which sound exciting. But no matter; I shall still look forward each year to my old friends, all of whom have their own stories to tell.

15

Some Pioneer American Entomologists

As many questions as there are still to be answered, we have nevertheless come a long way since the early settlers in the New World were awed by the strange plants and animals that surrounded them. Reports reached Europe of mosquitoes thick as the fogs of London, giant spiders, tiny birds that flew like hawk moths, snakes that rattled, plants that produced painful blisters. Europeans were keen to find new plants for their parks and gardens, new species of plants and animals to describe and catalog. The first step in knowledge is to know what one is dealing with, to have names for everything, to be able to group similar objects. The settling of North America coincided with the time of the earliest cataloguers of nature: John Ray in England and Joseph Tournefort in France. Within two years of the death of those primal naturalists, in 1707, Carl Linnaeus was born in Sweden and the Compte de Buffon In France. In 1749 the first part of Buffon's great *Histoire Naturelle* appeared, and Linnaeus was already preparing the fourth edition of his *Systema Naturae*.

It was a year before that, in 1748, that Linnaeus commissioned his student Peter Kalm to travel about the American colonies and send examples of as many species as he could back to his mentor. Kalm's *Travels in North America* was widely read, filled as it was with stories (often hearsay) about the flora and fauna of eastern North America. Even earlier, Mark Catesby's several years of travel had resulted in his *Natural History of Carolina, Florida, and the Bahama Islands* (1731), illustrated with his own paintings of plants, birds, fish, and mammals—as well as twenty-seven insects, including the luna and cecropia moths and the monarch butterfly.

But Kalm and Catesby each returned to Europe after a few years of wandering in the New World. John Abbot came to stay, to serve in the Revolution, and to settle in Georgia on land granted to him for his military service. As a youth in England, Abbot had collected insects and painted insects and birds. Bored with his job as a clerk and excited by his reading of Mark Catesby's *Natural History*, he set out for the colonies in 1773. After settling in Georgia, he did a little farming but spent most of his time collecting insects and painting them, often portraying the larvae and food plants along with the adults. Many of these he sold for a pittance to John Francillon, a silversmith of London, who sold them to others at a profit, without acknowledging their source except that they were from Georgia.

In the course of his long life, Abbot produced thousands of drawings and paintings, many of which found their way to museums in Europe. Sir James Edward Smith, the wealthy Englishman who had earlier bought Linnaeus's library and collections and carted them off to England, purchased many paintings of Lepidoptera from Francillon. In 1797 he brought out, at his own expense, two large volumes titled *The Natural History of the Rarer Lepidopterous Insects of Georgia Collected from the Observations of John Abbot, with the Plants on Which They Feed*.

So obscure a life did Abbot lead that he did not hear of this book until long after it had been published. Evidently he had no craving for recognition, and was content to spend his days collecting, painting, and shipping off his wares to patrons for just enough money to support his modest lifestyle. He is described by Robert Percy Dow as "an untutored optimist, with a constitutional smile, who looked forward only to the day's reward, who had talent with the brush, who had the assiduity to rear every insect species he could for fifty years." A friend of Abbot's, August Oemler, reported that in his old age Abbot was "still in the simplicity of a school boy," but deaf and so nearly blind and

---

The tiger beetle Cicindela formosa *is one of many insects discovered and named by Thomas Say.*

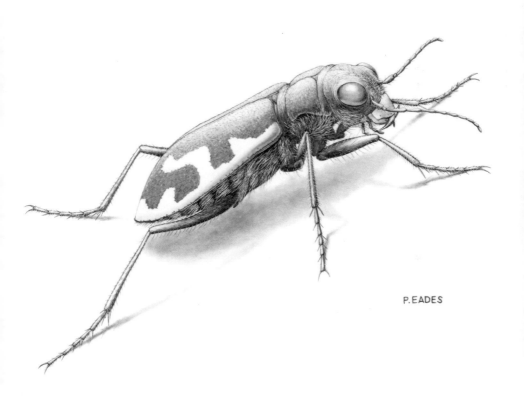

P. EADES

corpulent that he had to hire boys to catch butterflies for him. Nevertheless he was "cheerful, and his constitutional smile never left his countenance."

It was Oemler who had introduced Abbot to Linnaeus's *Systema Naturae*, which he read with astonishment. Abbot had been painting flowers all his life, without paying attention to the number of stamens. After studying Linnaeus, he was more careful!

Abbot was, however, not totally unknown to his contemporaries. He supplied specimens to Major John E. LeConte, a beetle enthusiast whose son, J. L. LeConte, became a noted student of beetles. Abbot corresponded with the Reverend T. M. Harris, author of *The Natural History of the Bible*, and offered to supply specimens to his son, Thaddeus W. Harris, about whom we will have more to say in the next chapter. He was even known to Thomas Jefferson, who suggested to pioneer ornithologist Alexander Wilson that Abbot might prove a source of Georgia birds. Wilson visited Abbot and commissioned him to collect birds for him, paying him forty-five dollars a specimen, far more than Abbot was used to receiving. Wilson made frequent references to Abbot in his *American Ornithology*.

That Jefferson knew of John Abbot is not surprising, for these were times when persons of power and influence were also persons of enlightenment. It was Jefferson, along with John Adams and Benjamin Franklin, who attacked the comments in Buffon's *Histoire Naturelle* that North American animals are smaller and more timid than those of Europe, living as they did under a "niggardly sky" and in an "unprolific land." When he was ambassador to France, Jefferson presented Buffon with a stuffed moose, forcing the latter to recant, though he never did so in print.

Franklin's interests in natural science were even broader than Jefferson's. His essay on mayflies, "The Ephemera," is one of his best known literary efforts. Franklin's inquiring mind led him to make one of the earliest studies of communication among ants. Finding ants in a pot of molasses in his house, he shook out all but one and hung the pot from the ceiling by a string. The one ant made her way up the string and down the wall to the floor. Soon a column of ants followed her course up the wall and down the string to the molasses, where successive waves of ants gorged themselves until the molasses was exhausted. Clearly the pioneer ant had carried a message to her nestmates.

Franklin's thoughts on entomologists were conveyed in a letter he wrote in 1760 to a young friend of his, Mary Stevenson:

> *Superficial minds are apt to despise those who make that part of creation their study, as mere triflers; but certainly the world has been much obliged to them. Under the care and management of man, the labours of the little silkworm afford employment and subsistence to thousands of families, and become an immense article of commerce. The bee, too, yields us its delicious honey. . . . A thorough acquaintance with the nature of these little creatures, may also enable mankind to prevent the increase of such as are noxious or secure us against the mischiefs they occasion.*

Both Franklin and Jefferson were well acquainted with America's first indigenous naturalist, John Bartram. A Pennsylvania farmer, Bartram had begun sending plants and natural history notes to his fellow Quaker Peter Collinson, in London, as early as 1734. In 1749 Collinson conveyed Bartram's observations on the "great black wasp of Pennsylvania" *(Sphex pensylvanicus)* to the Royal Society of London. Bartram's accurate observations stand as the first article on an insect written by a native of the New World. Later in his life, Bartram supplied specimens to various European scientists, including Linnaeus and Frederick the Great.

The person who, perhaps more than any other, deserves to be called the first American entomologist was also a Pennsylvanian, though in this case not a Quaker but a Lutheran minister. Frederick Valentine Melsheimer was born in Germany and reached the New World in a curious way. He was a chaplain in a regiment of Hessian mercenaries employed by the British in the Revolutionary War. Captured by American troops in the Battle of Bennington, Melsheimer was later paroled and sent to Pennsylvania, where he married a local girl and spent his life as a clergyman. He doubtless provided a source of amusement as he visited his parishioners carrying an insect net, for he was an ardent collector. In 1806 he published the first book on American insects: *A Catalog of Insects of Pennsylvania.* The catalog listed 1,363 species and included data on life histories and even a few recommendations on insect control. However, it is doubtful that farmers and gardeners made much use of his suggestions, since they were written in Latin. This historically important catalog has recently been reproduced in facsimile by the Entomological Society of Pennsylvania.

Two of Melsheimer's eleven children were also keen students of insects. John F. Melsheimer, a minister like his father, carried on an extensive correspondence with Thomas Say. Frederick E. Melsheimer became a country physician but carried on the family tradition, in 1853 publishing a *Catalog of the Described Coleoptera of the United States.* Louis Agassiz visited F. E. Melsheimer in 1860 and purchased the family collections for Harvard's Museum of Comparative Zoology, paying $150 for some 14,000 specimens.

Though living in rural Pennsylvania, the Melsheimers became involved in the intellectual ferment of Philadelphia in the late 1700s and early 1800s. Frederick V. was elected to the American Philosophical Society in 1795, the year that Jefferson was president of that society (which had been founded by Franklin in 1743, with John Bartram one of its charter members). Frederick E. Melsheimer was elected first president of the American Entomological Society in 1853.

Philadelphia had meanwhile become the site of America's first natural history museum. Founded in 1786 by artist Charles Wilson Peale in his own house, the museum contained a variety of animals, many of them prepared for exhibit by Peale himself. Franklin even contributed his angora cat and George Washington one of his golden pheasants (after both animals had died a natural death, of course). Later it boasted the skeleton of a mastodon, a three-headed snake, and bighorn sheep and pronghorns brought from the West by the Lewis and Clark Expedition. Peale's house soon became too small, and he was offered space in the new quarters of the American Philosophical Society. To save expense and to put on a show for the public, Peale hired several boys to carry the exhibits through the streets to their new home. As he described it "at the head was carried the American buffalo, then followed the panthers, tiger cats and a long string of animals of smaller size. The parade brought all the inhabitants to their doors and windows to see the cavalcade."

Peale's museum contained several thousand insect specimens as well as a large collection of shells. These were of special interest to a young native Philadelphian who worked in his father's drugstore but stole away when he could to haunt the museum. He came by his interests naturally, for Thomas Say was a great grandson of John Bartram. John was by this time dead, but his son William, Say's great uncle, was still tending the Bartram gardens and did much to encourage Thomas's interests in natural history. William Bartram's *Travels* had been published in 1791 and acquired a wide readership; Carlyle, Coleridge, and Wordsworth were among its admirers. Through his great uncle, and through his contacts at Peale's museum, Say became acquainted with Alexander Wilson, with botanist Thomas Nuttall, and with other naturalists. Nuttall and Say often worked at the museum late into the night, sometimes sleeping in the only space roomy enough to accommodate them: beneath the mastodon skeleton.

A group of Philadelphians of similar interests came to meet regularly at Mercer's Cake Shoppe, on Market Street, where on March 12, 1812, they founded the Academy of Natural Sciences. Thomas Say was evidently not present at this meeting, but in a subsequent meeting he was declared a charter member anyway and appointed "conservator," in charge of the library and collections. The latter consisted at first of "some half dozen common insects, a few madrepores and shells, a dried toad fish and a stuffed monkey." But that was soon rectified, and when the Academy obtained quarters at 35 Arch Street, Say became a regular inhabitant.

He had by this time failed in the drug business, and essentially lived in the Academy's rooms, surviving on the simplest of foods. Here, in 1816, he conceived of starting a book similar to the *American Ornithology* of his friend Alexander Wilson. It was to be a treatment of insect species "somewhat indiscriminately described" but published on unnumbered pages, so that they could eventually be bound "agreeably to systematic order" (as he stated in the Preface). For the colored plates, Say availed himself of the services of several artists, including Titian Peale, youngest son of Charles Wilson Peale (whose other sons were named Raphaelle, Rembrandt, and Rubens—all of them artists!). Titian and Thomas were to spend many months together in the field. *American Entomology* was never completed, as a result of diverse twists of fate in Say's life, but twenty-five years after his death J. L. LeConte edited a two-volume set of Say's complete writing on insects.

Thomas Say's first publications appeared in 1817 in the *Journal* of the Academy of Natural Sciences, which had been founded at the behest of its president, William Maclure, who had made a fortune in the English textile mills. In that year, Maclure, Say, and Titian Peale undertook a collecting trip to Florida, a major undertaking at that time. The trip was largely unsuccessful and was terminated abruptly because of the danger of Indian attacks. Say wrote to John F. Melsheimer about some of his adventures:

> *As we redescended the [St. John's] river we heard of parties of Indians who had been committing depredations, & one person informed us that a few days previous, his plantation was totally destroyed by them & his son killed. He narrowly escaped with the remainder of his family, & with the graze of a rifle ball on his forehead. The Indians then took the road to Picolata; so that we departed from that place in good time, as it seems probable they went in quest of us.*

Say did not long remain in Philadelphia after his return from Florida. Secretary of War John C. Calhoun wished to expand American influence in the western parts of the recently acquired Louisiana Purchase, and Major Stephen Long was appointed to head a scientific expedition to the sources of several western rivers. Disappointed with the results of his collecting trip to Florida and eager to obtain specimens from the unexplored West, Say applied for the position of expedition zoologist and was accepted. As he wrote to Melsheimer on March 13, 1819:

> *Our destination [is] to examine all the immense Western waters—the Mississippi, its tributaries, some of the Lakes & perhaps some of the rivers still further south. Besides my own department, viz. zoology, we shall be accompanied by a Botanist (Dr. Baldwin), a Geologist &*

Minerologist (Mr. Jessup), & Mr. T. Peale will accompany me to prepare the skins of such animals as may be discovered. . . . Our steamboat at Pittsburgh is nearly ready.

Long's steamboat, of unique design, got only as far as what is now Omaha, Nebraska, before winter set in—indeed it got no farther, and the expedition proceeded by foot and horseback thereafter. During the winter they remained in camp, studying the local Indians and collecting whatever animal specimens they could. Titian Peale was desperate to capture one of the "wolves," but the wily beasts consistently stole the bait without being captured. Eventually he got one and turned it over to Say for formal study. Thus, Thomas Say is credited with the description and naming of *Canis latrans,* the coyote, as he is with that of several other mammals and birds.

Say also earned a place in the annals of ethnology as a result of his descriptions of the customs and languages of the various Indian tribes they encountered. Ultimately he obtained many insect specimens that were duly described in the second section of his *American Entomology,* published in 1825. One of these was the notorious Colorado potato beetle, which we discussed in the previous chapter. Another was one of North America's larger and more ornate tiger beetles, which he named *Cicindela formosa,* "formosa" being the Latin word for "beautiful." Although Say's descriptions are simple and couched in a dry, straightforward style, one can now and then glimpse some of the excitement he must have felt in their discovery. *Cicindela formosa* is described in part as follows:

Red cupreous, brilliant . . . body beneath green or purple-blue, very hairy; thighs blue, tibia green. A beautiful species; it was captured by Thomas Nuttall, on the sandy alluvions of the Missouri river, above the confluence of the Platte.

In the final analysis, Long's expedition was not a great success. They failed to find the source of the Platte, the Arkansas, or the Red rivers, and the party struggled back half starved and without most of their horses. Many of Say's journals were lost when a group of deserters made off with much of their equipment. Long's descriptions of Colorado were also disappointing. The area was said to be "almost wholly unfit for cultivation," the lack of water providing "an insuperable obstacle in the way of settling the country." This was not what his superiors wanted to hear!

The expedition did, however, reach the Rockies, sighting what is now called Long's Peak, which they did not climb, though a few days later several in the party did ascend Pike's Peak. Say was ill off and on, as he was for a good part of his life. But the information he gleaned on the fauna and on the Indians eventually provided one of the most valuable parts of the report of the Long Expedition. And of course he was able to collect a good many insects.

Back in Philadelphia, Say was faced with the task of sorting, arranging, and describing all this new material. But he was soon off on another expedition with Long, this time into Minnesota and on into Canada. This trip, too, was filled with various adventures and misadventures, but as before Say was able to make substantial contributions to knowledge of the natural history and of the native tribes of the area.

Back at the Academy once again, Say immersed himself in preparing another volume of his *American Entomology.* About this time he began to correspond and exchange specimens with T. W. Harris, in Massachusetts. But soon he again yielded to temptation, or perhaps it was to a sense of obligation, for William Maclure had become involved in a scheme to set up a utopian community where his liberal beliefs on education could be realized. Say had, in fact, become financially dependent upon Maclure, as he remained for the rest of his life. Another wealthy British visionary, Robert Owen, had arrived in the United States a few years earlier. After discussions with Jefferson, John Adams, and others, he purchased the Rappite village of Harmonie, Indiana. Together Owen and Maclure recruited as many Philadelphia intellectuals as they could to enrich the community, to be called New Harmony. Thomas Say was on the "boatload of knowledge" that departed from Pittsburgh in December 1825. He was never to return to Philadelphia.

In New Harmony, Maclure founded the first kindergarten in the United States as well as the first free, coeducational elementary school. All possessions were to be held in common, and everyone was to participate in tilling the soil and in producing the necessities of life. Say's hands became "covered with hard lumps and blisters, occasioned by the unusual labor he was obliged to undertake." Within two years the dream of Utopia had collapsed, and Owen and Maclure had quarreled and departed. Say was left with even more responsibilities and distractions from his scientific work.

But there were brighter moments. Say began to put together his studies of shells, to be published as *American Conchology*. His artist was Lucy Sistaire, who had arrived on the "boatload of knowledge" and was described as the "handsomest and most polished of the female world" at New Harmony. On January 4, 1827, Lucy and Thomas eloped across the border to Illinois. Lucy was twenty-six, Thomas forty. Lucy urged her husband to move east, where they would be less isolated and have fewer distractions. But perhaps out of loyalty to the community, Say could not be moved. He continued to be periodically ill, and in 1834 he succumbed to a fever, his life's work far from complete.

Lucy maintained an interest in entomology and conchology and corresponded frequently with T. W. Harris, sending him copies of Thomas's manuscripts. As she explained to Harris: "Mr. Say was somewhat peculiar in writing his descriptions upon small pieces of paper, and . . . was not even particular in the choice of these." In 1841 Lucy Say was elected the first woman member of the Academy of Natural Sciences, where she deposited Thomas's library and collections. She outlived her husband by more than half a century, dying in Massachusetts in 1886.

What was Thomas Say like? He was described as warm, gentle, and cultured, as able to talk to trappers or Indians as to Philadephia intellectuals. He had "a passion for discovery" but little interest in mundane affairs such as money. Shortly after his death, his friend George Ord spoke of him to the American Philosophical Society:

> *The readiness with which Mr. Say attended to the wants of others, his liberality in communicating his knowledge to those who sought it, together with his urbanity and companionable qualities, were the*

occasion of such repeated interruptions, that he felt constrained to appropriate those hours to his private studies, which ought to have been devoted to rest; hence to him the season of midnight was the hour of prime, it was the time of stillness and tranquility; and so greatly did he enjoy these vigils, that he not unfrequently prolonged them, even during the summer, until the approach of day.

His disposition was so truly amiable, his manners were so bland and conciliating, that no one, after having once formed his acquaintance, could cease to esteem him. A remarkable feature in his character was his modesty, which, leading to habits of retirement, in some respects unfitted him for the intercourse of society, except that of his private friends, where, it may be said, he was truly at home, and where he was the idol of every heart.

In all, Say described over 1,500 species of insects. Unfortunately, his collections were ravaged by dermestid beetles, especially while they were in New Harmony. After they had been placed in the Academy, they were sent for study to T. W. Harris, in Cambridge. Harris found them to be "in a deplorable condition, most of the pins having become loose, the labels detached, & the insects themselves without heads, antennae, & legs, or devoured by destructive larvae & ground to powder by the perilous shakings which they have received in their transportation from Harmony. This irremediable destruction has in great measure defeated my expectation of deriving benefit from examining the specimens. . . ."

As a result of the brevity of Say's descriptions and the loss of his collections, the identity of many of the species he described has remained a matter of conjecture. A sad fate for a man who struggled so diligently to put American entomology on a sound footing! One wonders how matters might have turned out had he not been lured to New Harmony. Perhaps Say's finest hours were those when, as a young man on Long's expeditions, he recorded the languages and customs of the Indians and collected insects, birds, and mammals in the virtually unknown western states. The inscription on the monument to Thomas Say's memory, at New Harmony, is an especially fitting one:

Votary of Nature even from a child,
He sought her presence in the trackless wild;
To him the shell, the insect, and the flower
Were bright and cherished emblems of her power.
In her he saw a spirit all divine,
And worshipped like a pilgrim at her shrine.

16

More Early American Entomologists

Soon after the American Revolution, about the time John Abbot was settling in Georgia and Frederick Valentine Melsheimer in Pennsylvania, a young man was beachcombing along the coast of Maine. He was at loose ends, for his mother had died and his father had left his trade in Boston as a naval architect and had retired to a small farm in Kittery. William Dandridge Peck had graduated from Harvard in 1782, but after a brief career in a counting house he joined his father in Kittery, where he became something of a recluse. One day he came upon a shipwrecked vessel, and among its remains was a copy of a book he had not encountered in his career at Harvard. It was *Systema Naturae,* by the Swedish naturalist Carl Linnaeus—the same book that John Abbot learned of only late in his life. In Peck's case, it gave direction to his young life: he began to study the birds, fishes, and insects of the Maine coast.

Peck's collection of insects grew rapidly, and he built a crude microscope so that he could better study them. He exchanged specimens with William Kirby in England. In 1895 his first paper appeared in the *Massachusetts Magazine:* "The description and history of the cankerworm." Unlike Thomas Say (who was still a child at the time), Peck did not take it upon himself to extend the Linnaean system to American insects. He was mainly concerned with learning more about the life histories of the species he encountered, especially those that were attracting attention because of their depredations. For his paper on the cankerworm (later reprinted in *The New England Farmer*) he received an award of fifty dollars and a gold medal from the Massachusetts Society for Promoting Agriculture.

The Society was at the time raising funds to create a professorship of natural history and a botanical garden at Harvard. Francis Cabot Lowell, the scion of two old Boston families, was in charge of the subscription list. It appears that the Society had Peck in mind from the first. His father and mother were from respected Boston families, and William had earned a reputation not only for his several articles on insects, fish, and plants but for his knowledge of classical literature, painting, and architecture. Peck accepted the professorship with some reluctance, however, as he was a modest man and had lived something of a hermit's life up to that time.

Since part of his task was to establish a botanical garden, he was sent to Europe to study gardens there and to obtain seeds, plants, and books. In London he met Sir Joseph Banks and visited his recently founded gardens at Kew. He also met his old correspondent William Kirby as well as Sir James Edward Smith, who had published John Abbot's plates of Georgia insects a few years earlier. Peck also visited France, Germany, the Netherlands, and Denmark, but most of his time was spent in Sweden, where he met some of Linnaeus's students. He remained in Europe three years, shipping back many books and specimens and engaging a gardener, William Carter, who remained with the Harvard botanical gardens for many years. In the words of Jeannette Graustein, in her biography of Thomas Nuttall (who succeeded Peck in the professorship), in Europe Peck's "quiet worth made valuable friends not only for himself but also for Harvard and the struggling natural sciences in America." Many of these friends corresponded with Peck for the remainder of his life.

Back at Harvard, Peck began a course of instruction in natural history. His students included several whose fame eventually exceeded that of their teacher: Charles Pickering, said to be the most versatile naturalist of his time; Ralph Waldo Emerson, whose early poems and journals are sprinkled with bits of natural history learned from Peck; and Thaddeus William Harris, who extended the interests of his mentor to become America's first applied entomologist.

In 1810 Peck married Harriet Hilliard, the daughter of a clergyman, and the pair settled in the Garden House, built for them in the botanical gardens. Peck was a fellow of the American Academy of Arts and Sciences, a member of the American Philosophical Society, and a founder of the American Antiquarian Society. He continued to teach and to supervise the gardens until his death in 1822, publishing occasional papers on insects, chiefly those that attack shade, forest, and fruit trees.

Because of Peck's modesty and his somewhat retiring nature, he achieved little lasting fame. As his obituary, in the Collections of the Massachusetts Historical Society, somewhat quaintly puts it:

> *There was nothing about Mr. Peck's life or character which could furnish the materials of a highly wrought picture; nothing which would address itself to the passions or the imagination. It was simply the example of an unaided and retired individual, struggling, during the greater part of his life, against every discouragement, upborne by his genius and love of study, and constantly adding new stores to a powerful mind, capable of comprehending all that it received from reading and observation, and of analyzing, arranging and preserving it.*

T. W. Harris later said of Peck: "It was this early and much esteemed friend who first developed my taste for entomology, and stimulated me to cultivate it." Thaddeus's mind may, however, already have been primed for this response, for his father, the Reverend T. M. Harris, had written the *Natural History of the Bible,* in which every plant and animal in the Bible had been described and identified. Furthermore, his mother had reared silkworms and used the silk in making clothing.

Thaddeus William Harris was born in Dorchester, Masschusetts, and graduated from Harvard in 1815. He remained at Harvard another five years and obtained a medical degree. For eleven years he practiced medicine in communities near Boston, but he seemed to regard medicine as a distraction from his real interest, the study of insects. While sharing a practice with Dr. Amos Holbrook he did, however, acquire a wife, the doctor's daughter Catherine. The marriage proved a long and happy one.

Their first child was born in 1826, their twelfth and last child in 1849.

In the effort to escape from a profession he did not enjoy, Harris applied for an appointment as librarian at Harvard, a position his father had held for a few years much earlier. He was passed over, but five years later the position became vacant again, and this time he was appointed. He held this position for the remaining twenty-five years of his life. Although he had looked forward to the librarianship, it proved more onerous and time-consuming than he had supposed. He had hoped that it might lead to a professorship in natural history similar to that held by Peck some years earlier, but the position went unfilled for lack of funds until 1842, when Harris was passed over in favor of Asa Gray, the eminent botanist and later the defender of Darwin against the attacks of Louis Agassiz.

In the meantime Harris did, however, have an opportunity to lecture to the Boston Natural History Society and to give some instruction in natural history at Harvard. One of his students was Thomas Wentworth Higginson, who later became a liberal minister, an ardent abolitionist, and a discoverer of new talent, including Emily Dickinson. In a *Memoir* of Harris, Higginson described Harris's course as follows:

> *There were exercises twice a week . . . with occasional elucidations and familiar lectures by Dr. Harris. . . . This was the only foothold which Natural History had then secured in what we hopefully called the 'university.' Even these scanty lessons were, if I rightly remember, a voluntary affair; we had no 'marks' for attendance, and no demerits for absence, and they were thus to a merely ambitious student a waste of time, so far as college rank was concerned. Still they proved so interesting that Dr. Harris formed, in addition, a private class in entomology, to which I also belonged. . . . These were very delightful exercises, according to my recollection. . . . Dr. Harris was so simple and eager, his tall, spare form and thin face took on such a glow and freshness, he dwelt so lovingly on antennae and tarsi, and handled so fondly his little insect-martyrs, that it was enough to make one love this study for life, beyond all branches of Natural Science, and I am sure it had that effect on me.*

Although Higginson went on to become a major figure in the literary "flowering of New England," indeed he never lost his love of natural history and especially of insects. In an essay, "Butterflies in Poetry," written when he was much older, Higginson wrote of

> . . . *the eternal youthfulness of Nature [that] answers to my own feeling of youth and preserves it. As I turn from these men and women around me, whom I watch gradually submerged under the tide of gray hairs—it seems a bliss I have never earned, to find bird, insect and flower renewing itself each year in fresh eternal beauty, the same as in my earliest childhood. The little red butterflies have not changed a streak of black on their busy wings, nor the azure dragonflies lost or gained a shade of color, since we Cambridge children caught them in our childish hands. Yesterday by a lonely oak grove there fluttered out a great purple butterfly. . . . It was the beautiful* Papilio philenor, *which Dr. Harris showed us in college.*

Harris's first publication appeared in *The New England Farmer* in 1828, while he was still a practicing physician. Many others followed, chiefly in this and other agricultural journals. Harris did not have a background in agriculture, but he received payment for these articles, and considering his meager salary from Harvard and his growing family, such payment was surely welcome. Furthermore, he was looked upon as the only ready source of information on insects of concern to gardeners. Higginson remarks that "many an old farmer who had travelled miles to bring him a newly discovered cabbage-pest, or a strange wheat-fly, was sent away from his house delighted with the story of insect life, and the practical hints that he received from the reserved but courteous gentleman who had welcomed him with the dignity and politeness of a by-gone age."

In 1833 the Commonwealth of Massachusetts published a report on the geology, botany, and zoology of the state, and to it Harris appended a catalog of insects—some 2,300 species, a remarkable number for a person to have identified who was an entomologist neither by training or profession. As his reward, Harris received a single copy of his catalog. A few years later, a state commission was appointed to prepare a more thorough survey of the botany and zoology, and Harris was asked to cover the insects. The result was his *Report on Insects Injurious to Vegetation,* published in 1841 and republished in 1842 as a *Treatise.* He was paid $175 for completing a book that has become a classic of entomology. In 1862, some years after his death, the State Board of Agriculture published an expanded, extensively illustrated edition.

The *Treatise* contains a wealth of information on the life histories of insects, much of it based on Harris's own observation but much of it also gleaned from his reading and from his correspondence and personal contacts. Much of this information is still valid, although this is hardly true of his suggestions concerning control. "Hand picking" and "shaking from the tree" were common recommendations, and for "hairy caterpillars": "pay children to collect them by the quart." Insecticides included whale-oil soap, ashes, lime, and decoctions of tobacco leaves. Harris took seriously his charge from the governor:

> It is presumed to have been a leading object of the Legislature, in authorizing this Survey, to promote the agricultural benefit of the Commonwealth, and you will keep carefully in view the economic relations of every subject of your inquiry. . . . That which is practically useful will receive a proportionally greater share of attention, than that which is merely curious; the promotion of comfort and happiness being the great human end of all science.

This brought a blast from Concord sage Henry David Thoreau, who in characteristic manner took it out on his *Journal,* May 1, 1859:

> We accuse savages of worshipping only the bad spirit, or devil, though they may distinguish both a good and a bad; but they regard only that one which they fear and worship the devil only. We too are savages in this, doing precisely the same thing. This occurred to me yesterday as I sat in the woods admiring the beauty of the blue butterfly. We are not chiefly interested in birds and insects, for example, as they are ornamental to the earth and cheering to men, but we spare the lives of the former only on condition that they eat more grubs than they do cherries, and the only account of the insects which the State encour-

ages is of the 'insects Injurious *to Vegetation'. . . .*
Children are attracted to the beauty of butterflies, but their
parents and legislators deem it an idle pursuit. The parents remind me
of the devil, but the children of God. Though God may have pro-
nounced his work good, we ask, "Is it not poisonous?"

Thoreau and Harris were, however, the best of friends, and
Thoreau (twenty-two years younger than Harris) often sought
the advice of the Harvard librarian on insect problems. In his
Journal, for example, he records the following:

On June 19, 1854: "Went to Cambridge. . . . Dr. Harris says that my
cocoons found in December are of Attacus cecropia, *the largest of*
our emperor moths. He made [drawings] of the four kinds of emperor
moths which he says we have. . . . In a large and splendid work on the
insects of Georgia, by Edwards and Smith (?), near the end of the last
century, up-stairs, I found plates of the above moths. . . . [This book
was, of course, the little-acknowledged work of John Abbot].
June 16, 1855: "Carried to Harris the worms—brown, light-striped
and fuzzy black caterpillars . . . also two black beetles; all of which I
have found within a week or two on ice and snow.

Thoreau had a well-worn copy of Kirby and Spence's *Intro-*
duction to Entomology as well as Harris's *Treatise*. Harris sometimes
asked him to collect specimens for him. For example, on June
27, 1854, Harris wrote to Thoreau:

Your letter of the 25th, the books, and the Cicada came to hand this
evening,—and I am much obliged to you for all of them . . . for the
letter, because it gives me interesting facts concerning Cicadas; and for
the specimen because it is new *to me, as a species or as a variety. . . . I*
should be very glad to get more specimens and of both sexes. Will you
try for them?

All of which led Harris to comment to Bronson Alcott that if
Emerson had not spoiled him, Thoreau would have made a
good naturalist.
Harris corresponded for a time with John Abbot, with F. E.
Melsheimer, and with Thomas Say; after Say's death he studied

the remnants of his collection, as we have noted. He was a life-long friend and correspondent of Nicholas Marcellus Hentz, a pioneer student of spiders, who like Harris was an amateur in the sense that he earned his living in another field, in this case as a language teacher. Foreign correspondence included two distinguished English entomologists, William Kirby and Edward Doubleday. Doubleday, a lepidopterist of the British Museum, is best remembered as the inventor of the practice of sugaring for moths. Harris had a visit from Doubleday, and was so taken by him that he named one of his sons after him. In a letter to Doubleday in 1839, Harris expressed some of the frustrations he felt because of his isolation from others of like interests:

> *You have never, and can never know what it is to be alone in your pursuits, to want the sympathy and the aid and counsel of kindred spirits; you are not compelled to pursue science as it were by stealth, and to feel all the time, while so employed, that you are exposing yourself, if discovered, to the ridicule, perhaps, at least contempt, of those who cannot perceive in such pursuits any practical and useful results. But such has been my lot, —and you can therefore form some idea how grateful to my feelings must be the privilege of an interchange of views and communication with the more favored votaries of science in another land.*

Harris's insect collection was large and in far better condition than Say's, "not only the best," said Harris, "but the only general collection of North American insects in this country." Harris had invented a device for keeping his collection free of dermestid beetles. It was a steel chamber connected with a steam pipe, such that boxes of insects could be placed in it and subjected to temperatures high enough to kill the pests. This worked fine except on one occasion when he received a shipment of butterflies pinned in boxes with bottoms of beeswax rather than cork. Harris's remarks when he removed the molten mess from the chamber have not been preserved.

As Harris grew older and the Harvard library grew in size,

Cicadas were among the insects collected for T. W. Harris by Concord sage Henry Thoreau.

P. EADES

he found that he had less and less time to devote to insects, although he still used most summer weekends and vacations for studies out-of-doors. Higginson describes Harris's field trips as follows:

> *His excursions . . . though rare, were effectual; he had the quick step, the roving eye and the prompt fingers of a born naturalist; he could convert his umbrella into a net, and his hat into a collecting-box; he prolonged his quest into the night with a lantern, and into November by searching beneath the bark of trees. Every great discovery was an occasion for enthusiasm.*

Harris looked forward to retirement. "When that time does come [he wrote a friend] I mean to go forth into the neglected parts of this and neighboring States, and collect largely of the insect treasures contained in them." But it was not to be. In 1844 he wrote to Doubleday:

> *My friend, be warned in season, if it be not now too late, by my own sad experience, that the unremitting devotion to duties in a public establishment will wear out body and spirit, will deprive you of leisure, of necessary exercise and relaxation, and will give you in return only a petty compensation, at the expense of your time, health and happiness.*

In the following year, Harris had an attack of pleurisy, and on January 16, 1856, just after discussing library affairs with his assistant, he succumbed to a heart attack. A month later a tribute was paid to Harris at a meeting of the American Academy of Arts and Sciences, speaking of his "singular facility in employing language intelligible to the common reader, and at the same time fulfilling all the requirements of science, which render him a model for the interrogator of Nature." Louis Agassiz seconded the tribute and added that Harris had "few equals, even if the past were included in the comparison."

But a new generation was ready to take the stage, and because of the influence of Harris's *Treatise,* they could rightfully call themselves "entomologists." When the 1862 edition of the *Treatise* appeared, Agassiz presented copies to all his students, two of whom became major figures in the further development of the science. These were Alpheus Spring Packard and Samuel Hubbard Scudder,* both of whom (with others) broke away from Agassiz in 1863 because they were not allowed to publish or to maintain a personal collection (they were later reconciled). (Tributes to Packard and Scudder were written by T. D. A. Cockerell, whom we shall meet in the next chapter.)

L. O. Howard, who was chief of the Federal Bureau of Entomology from 1894 to 1927, was given a copy of the *Treatise* as a Christmas present when he was fourteen. "I shall never forget my delight," he wrote. "I feel sure that it made many entomologists."

John Henry Comstock, who founded the Department of Entomology at Cornell University, came upon the *Treatise* in quite a different way. In 1870 he was twenty-one and a steward on a schooner carrying cargo on the Great Lakes. During a stop in Buffalo, New York, he spotted a copy of Harris's book in a bookstore. He did not have enough money to buy it, but the next day he drew some money from his pay from the captain and returned to the bookstore to buy it. Later he wrote on the flyleaf:

> *I purchased this book for ten dollars in Buffalo, N. Y., July 2, 1870. I think it was the first Entomological work I ever saw. Before seeing it I had never given Entomology a serious thought; from the time that I bought it I felt that I should like to make the study of insects my life work.*

We may as well let Harris have the last word (from his very first publication, in *The New England Farmer,* 1828):

> *An interest in the science of entomology appears to be awakened among us, and we are gradually becoming sensible of the utility of the*

---

* Scudder edited a collection of Harris's correspondence, published in 1869 by the Boston Society of Natural History. In the Preface he specifically states that he had never met Harris. This statement refutes that in J. S. Wade's essay on Harris and Thoreau (cited in References) that Scudder was a student in Harris's classes (this was repeated in E. O. Essig's *History of Entomology,* 1931). Scudder was, in fact, born in the same year that Harris began his classes, and was five years of age when Asa Gray took over the teaching of natural history.

pursuit. To mention nothing more, the frequent inquiries, made in public journals, respecting the economy and metamorphosis of insects, sufficiently indicate a desire for further information. There may be some who object to the study of insects on account of their apparent unimportance and insignificance. These may be assured that it is a never failing source of the most rational enjoyment; and that there is as much to be discovered and to astonish in magnifying an insect as a star.

17

A Pair of Latter-Day Entomologists

Entomologists are above all individualists. To a degree they live in a world apart from the masses that fail to appreciate the objects of their study. So intrigued are they with the bizarre forms and life histories of insects that human affairs and artifacts often seem paltry by comparison. So it is not surprising that they sometimes seem obsessed and idiosyncratic, that their lives sometimes take peculiar turns.

To trace the history of American entomology after the death of Say in 1834 and of Harris in 1856 would require a very fat volume—and to an extent Arnold Mallis has already filled that gap with his book *American Entomologists.* So in this chapter I shall merely pick two favorites, two particularly noteworthy specimens who, having crossed the meadow for the last time, should be pinned with their wings spread and displayed in a particularly prominent place.

Theodore Dru Alison Cockerell came from a distinguished English family. Several members of the family attained knighthoods; T. D. A., the renegade, published 3,904 articles, mostly on insects—135 in one year (1897), more than many scientists publish in a lifetime. According to William A. Weber, of the University of Colorado, who has compiled his bibliography and edited many of his letters, Cockerell "was obsessed with the drive to communicate to others the thoughts of his lively intellect, the facts he had learned, and to carry on dialogues on the great issues confronting the world."

Cockerell was not only an entomologist; he was broadly

concerned with the human endeavor, having been influenced in his youth by the British socialist William Morris. He dabbled in poetry and in art; as a scientist he published on snails, fish scales, the genetics of sunflowers, living and fossil plants, fossil insects, and living insects of several groups, but especially scale insects and bees. He never owned a typewriter, but scratched out his papers in a minuscule longhand and sent them off to publishers (of 221 different periodicals!) without much revision. Needless to say, he incurred a certain amount of criticism for his habit of churning out so many short papers on such a variety of subjects, often merely isolated descriptions of new species.

It has been said that Cockerell's precarious health led him to believe that he might die at any time, so that he must submit his thoughts to publishers as quickly as possible (he lived to be eighty-two!). There is not much truth in this; rather, he was an enthusiast for the living world, and a superb communicator. He was a renaissance figure who lived well into the twentieth century, a generalist in a world of specialists, a romantic who would scarcely survive in our modern world of science, ruled as it is by mathematics, by peer review, and often by those more concerned with their own egos than with the natural world.

T. D. A. Cockerell was born in Norwood, England, in 1866. His father introduced him to books on natural history at an early age, and a teacher, Sarah Marshall, encouraged him to collect specimens. When still a child, he drew up descriptions of the insects he collected and invented names for them. Joseph Green's *Insect Hunter's Companion* and Figuier's *Insect World* became his bibles. He was a sickly lad, and when he was twenty he was diagnosed as having tuberculosis. Fortunately, he had a well-to-do friend who was on his way to Colorado, the mecca of everyone with lung problems. Colorado was also a favorite destination of younger sons of English landowners, who traditionally left their estates to their eldest sons. Isabella Bird's *A Lady's Life in the Rocky Mountains* (1880) had been widely read in England. As a result the state was almost an outpost of England.

Cockerell and his companion settled in West Cliff (now called Westcliffe) in the Wet Mountain Valley southwest of Colorado Springs. The many letters he wrote from West Cliff tell of the simple, rustic life he led. Most of these letters were addressed to

Frederic Fenn, the brother of the girl he had left behind in England, Annie Fenn. Annie's father had little use for T. D. A.'s socialist leanings and considered him a health risk, so there was no opportunity to communicate directly with her. Hence the letters to Frederic, who turned them over to Annie. Annie destroyed them, but only after copying them into notebooks and clipping out Cockerell's delightful sketches. These eventually came into Cockerell's possession and found their way into the University of Colorado library. Many years later, Cockerell summed up this period in his life as follows:

> *The three years in Colorado during which I lived in Wet Mountain Valley were occupied with various pursuits, whereby I earned enough to keep me going, but always the major interest was natural history. I corresponded with many naturalists, and as early as 1887 began a catalogue of the entire Fauna and Flora of Colorado, recent and fossil. It included every reference I could find in the literature, very many other references kindly supplied by correspondents, and of course such of my own finds as I could get identified.*

It was characteristic of the man that after a few months in a new environment he set about to catalog the whole of its natural history! He also founded a Colorado Biological Association and solicited memberships from naturalists in many parts of the world.

In the fall of 1888, Cockerell and several companions crossed the Rockies by horseback and buckboard, going nearly to the western border of the state.

"I . . . have been collecting plants and insects," he wrote from a remote campsite to Frederic Fenn on September 15, "writing, and reading Darwin's *Earthworms.*" And on September 22: "I have spent most of the day in making a catalogue of the insects collected on this expedition; I did the Neuroptera some days ago; today I have done the Coleoptera, Lepidoptera and Orthoptera. I find it an excellent idea to go over the collection in this way, writing down locality and date and a short description of each. It fixes the different species in my mind. . . . "

Cockerell was an ardent evolutionist, and one of his correspondents was Alfred Russel Wallace, co-proposer with Darwin of the concept of evolution by natural selection. When Cockerell returned to England in 1890, Wallace employed him to edit the second edition of *Island Life*. Cockerell cherished his friendship with Wallace, which lasted until Wallace's death many years later, in 1913. The two discussed many topics, "all the debatable biological and sociological questions of the day," as Cockerell wrote later, adding that Wallace "believed in, and of course illustrated by his own conduct, the right of any man to study what he chose, and not be limited in his intellectual activities because his colleagues had labelled him this or that."

In 1891 Cockerell was appointed curator of the museum of the Institute of Jamaica, perhaps because of Wallace's influence. He had by this time married his childhood sweetheart, Annie Fenn, as he was now gainfully employed and ostensibly cured. However, in a tropical environment his health began to regress, and in 1893 he took advantage of an opportunity to move to the New Mexico Agricultural College. It was here that he first became interested in bees, being especially intrigued by a tiny yellow bee, *Perdita luteola*, which visited the yellow flowers of composites in great numbers. He began to study the genus *Perdita*, and soon became broadly involved in the bees of New Mexico, of the United States, and eventually of the world. "From this time onward," he wrote much later, "I have never ceased to work on bees, and have published 5,480 new names for species, subspecies, and varieties, and 146 names for genera and subgenera." (He went on to describe bees for another ten years after writing these words!)

Cockerell's health improved in New Mexico, but he was plagued by a series of family tragedies. In Jamaica, Annie had given birth to a son who died in a few days. Annie herself died soon after their arrival in New Mexico while giving birth to a second son, who was to live only a few years. But T. D. A.'s fortunes turned for the better when he met (while collecting *Perdita*) a new teacher of biology at the local high school, Wilmatte Porter. Wilmatte became his wife, collaborator, and close companion for the rest of his life.

After a short stay at New Mexico Normal College, at Las Vegas, and at Colorado College in Colorado Springs, the Cockerells moved to Boulder where, in 1906, he was appointed a professor at the University of Colorado (though himself without a college degree). He remained there until his retirement in 1934, though with many bee-hunting excursions to odd corners of the world: Honduras, Peru, Africa, Australia, Siam, and Japan, among others. Though admitting the need to study the biology of bees, he often evaluated his trips in terms of the number of new bees he collected: for example 457 new species in a 1931–32 expedition to Africa. Returning from Siberia in the wake of a major earthquake, Cockerell characteristically cabled home: "Bees safe, Theo."

In 1906 Cockerell spent much time in the fossil insect beds at Florissant, Colorado, along with his wife, a student, S. A. Rohwer, and Dr. William Morton Wheeler, then at the American Museum of Natural History in New York. These were exciting times.

"The opportunity for discoveries was too good to be missed," wrote Cockerell twenty years later, "so for several years we had Florissant Expeditions in the field, always productive of good results. My wife said that hunting fossil insects has all the fascination of gambling, with none of the attendant evils. Slab after slab would be turned up, usually with nothing to show, but once in a while a good specimen would appear, and the new species certainly averaged more than one a day. On one occasion my wife turned up a piece of shale in a new locality, and there was a butterfly, the spots still showing on the wing."

Later, Cockerell studied insects in the Green River shales of Wyoming as well as fossil beds in England, Argentina, and Siberia. But in 1937 he lamented: "I now have before me many more bees than I can ever hope to study, and must give my main attention to these, leaving the fossil insects to younger and I hope more competent workers."

Joseph Ewan, author of *Rocky Mountain Naturalists* (which he dedicated to Cockerell), visited the professor in his cluttered laboratory in Boulder in 1940. "In his seventy-fourth year," Ewan wrote, "Professor Cockerell was slender, as he had always been, of average height, soft-voiced and even a little sibilant of speech yet steady of finger and with little dimming of the wide blue eyes that his students remember so well. He usually dressed in a light grey suit and wore his coat and waistcoat in even the warmest weather invariably closed at the throat by a more unconventional ample dark green cravate of the sort commonly associated in this country with artists. He was crisp but unhurried in conversation; in movement purposeful."

Others have spoken of Cockerell's wit and his "charming whimsicality." As a professor, he tended to talk, rather than lecture, to his classes, often ranging over a wide variety of biological topics. He preferred to arouse a sense of excitement and exploration in his students rather than cramming them with facts. Both he and his wife often talked on natural history subjects to lay audiences.

Cockerell was always torn between his devotion to his objects of study, whether they were bees, fossil insects, scale insects, mollusks, or plants, and his desire to become involved in broad topics of biology or of human affairs. His bibliography is sprinkled with titles such as "What Science Can Do to Make the World a Better Place in which to Live," "Education and Human Nature," and, in 1932, "A Remedy for the Depression."

In 1916 Cockerell's talents as a poet, a teacher, and an evolutionist were blended into an amusing five-act play, *Progress, a Drama of Evolution,* in which the actors are oysters, sea urchins, frogs, dinosaurs, jackals, and others. It is too good to be forgotten! Jackal says, in Act IV:

> *Here comes a man, we'll call him to account;*
> *Let him excuse himself as best he may.*

To which man (clearly Cockerell!) replies a bit later:

> *Frail, imperfect, failing ever,*
> *Stumbling on till death may sever*
> *Chains that bind the soul:*
> *May heaven judge me by my meaning*
> *Striving, searching, ever gleaning*
> *Parts of nature's whole.*

In later years, Cockerell returned to a subject he had first explored with A. R. Wallace: island life. He chose the Channel Islands, off the coast of southern California, and after some difficulties managed to reach all the major islands, traveling by air for the first time at the age of seventy-two. He was still involved in the fauna, flora, and geology of the islands—and in describing bees—when he died in San Diego in 1948.

I was a graduate student when Cockerell died and never met him, but like many entomologists of my generation I had many indirect contacts with him. When writing a book on sand wasps, I had occasion to study a fossil he had described from Florissant, and when monographing a genus of wasps from Australia I had to consider several species he had described from that continent. In neither case did I find it necessary to challenge his conclusions.

Cockerell was a leading character in the biography of William Morton Wheeler that my wife, Mary Alice, and I published in 1970. I also count as friends several people who knew and were influenced by Cockerell. Charles Michener, of the University of Kansas, spent a summer with the Cockerells when they lived in Boulder, where the professor "always gave freely of his vast store of knowledge about bees." E. Gorton Linsley, of the University of California, knew Cockerell from his California days, and found him most generous with his time, his library, and his collections. In addition to his extensive publications, wrote Linsley, "his indirect contribution as a teacher, correspondent, and friend of other biologists has also been most significant."

T. D. A. Cockerell has been called, with some justification, "the last survivor of that illustrious assemblage of naturalists of the Victorian era." His joy of discovery was so irrepressible, his faith in science so complete that he sought to inform his colleagues abundantly and to educate those who were so unfortunate as to be unable to appreciate a bee, a mealybug, a piece of shale displaying an insect that lived thirty million years ago. His taxonomic work tended to be superficial (he had a microscope but used it only to check minor details, preferring the simplicity of a hand lens). His writings were often inadequately referenced—he had a phenomenal memory for everything he had read, and seemed to assume that his readers had a similar memory. He was rather too occupied with delineating "new species," with little regard for their life histories or the roles they played in nature. Like Thomas Say, a century earlier, he was over-whelmed by the need to fill in gaps in knowledge of recently discovered faunas. Like Say, too, he was largely self-taught, but through his reading and especially through contacts with other biologists, he became one of the more prominent figures of his time.

Fortunately, Cockerell was blessed with a long life, a life in which his energies never flagged, his enthusiasms never dimmed. He may have been an anachronism by the mid-twenti-eth century, because we live in a time when the contents of most scientific books and journals are comprehensible only to special-ists. This is a necessary stage in the history of science; but we cannot afford to lose a sense of wonder, a sense of surprise at the bounty and diversity of nature. The wide blue eyes and facile pen of T. D. A. Cockerell were needed; and we still need them.

Having decided that I would like to include in this series of portraits someone I had known personally, I was left with a dilemma. I have known a good many entomologists and have liked most of them. No matter how different their backgrounds and personalities, all have shared an awe of the diversity and intractability of the insect world. The fact that so many other people fail to share their admiration for insects renders them a particularly close-knit coterie.

I elected to discuss a person whom I learned to admire when quite young but did not meet until a few years before his death in 1967. Francis Xavier Williams was in many ways the antithesis of Cockerell, a man who was concerned with learning what insects *do*, and who described species only when it was neces-sary to have names for them. I am not aware that Williams ever met Cockerell (who was an older man by sixteen years), but Williams did rely on Cockerell to identify his bees and on Cock-erell's student S. A. Rohwer for help with his wasps. Unlike Cockerell, Williams left no autobiographical writings and few ruminations on human society. The number of his publications cannot rival Cockerell's (a mere 286!), but within their genre

---

Tiny yellow Perdita *bees first attracted T.D.A. Cockerell to this fascinating group of insects.*

P. EADES

they have rarely been surpassed.

F. X. Williams was brought up in the San Francisco area and, like so many incipient entomologists, began collecting insects when a small boy. His father encouraged him by having insect boxes made for him and by letting Francis take over an attic room as a laboratory. San Francisco in the 1890s still had much open space where a boy could wander and fill his collecting jars. Earlier, the area had been covered with dense growth of chaparral, but by this time most of it had been cleared except close to the coast, where there were cliffs and sand dunes. "As a consequence," Williams wrote in 1910, "the insects are disappearing along with the destruction of their food, and the hunting grounds of the entomologist [are] becoming more and more restricted, and are soon destined to become a thing of the past."

One of the butterflies Williams collected was the now-extinct xerces blue, which has become the symbol of the Xerces Society for the preservation of butterflies and other wildlife. In 1910 it was already "quite rare, though formerly an abundant insect." His youthful paper on the butterflies of San Francisco is filled with comments on the decline of various species, a decline that has, of course, now become a rout of the insect fauna. The same could be said of every major city, but the San Francisco area was blessed with a particularly distinctive fauna, especially that associated with the dunes along the coast and their unusual plant life.

Fortunately, Williams developed the habit of depositing the specimens he collected in major museums, where they would be preserved for posterity. Most of the material he collected in the San Francisco area at the turn of the century and later is now housed in the California Academy of Sciences, where it provides an invaluable record of a greatly diminished fauna. Williams became a member of the Academy in 1918, but he was associated with it informally much earlier. At the Academy he met Edwin C. VanDyke, a physician with an incurable enthusiasm for beetles, who later became a professor at the University of California. His younger colleague, E. O. Essig, reported that VanDyke was "an earnest, strenuous, untiring, careful and painstaking collector. . . . He beat trees and brush with consideration. If he turned over a stone or a log, he carefully replaced it . . . and saw to it that his students and associates followed his example." Though of "fearless demeanor," VanDyke was an inspiration to many a young entomologist.

The California Academy of Sciences was virtually destroyed during the great earthquake and fire of 1906, and only small parts of its library and collections could be saved. Williams was at this time far from San Francisco. While a student at Stanford, he had been invited the year before to join the Academy's expedition to the Galapagos Islands as entomologist. The islands had received no thorough faunal survey since they first became well known through Darwin's *Voyage of the Beagle*, and it was the purpose of the expedition "to make an exhaustive survey . . . before it proved too late." An eighty-nine-foot, thirty-year-old, two-masted schooner was purchased and rechristened the *Academy*. Sailing from San Francisco on July 3, 1905, the expedition reached the Galapagos on September 24, spending a full year there and visiting all the islands, often more than once. There were eleven persons aboard, eight scientists and three crew members. It must have been a heady experience for twenty-three-year-old Francis.

As the log of the voyage has since been published, we know a good deal about the expedition. Most of the party had not been to sea before, so it is no surprise that seasickness quickly became a problem. Standing watch while becalmed off Mexico, Williams caught a "black witch," the great moth *Erebus odora*. Off Cocos Island, he spotted a school of marine water striders, *Halobates*. Once at the Galapagos, the group went ashore nearly every day, and sometimes camped ashore, returning to the ship each time "laden with specimens." They often "lived off the land," shooting and eating the feral goats and pigs as well as devouring native fish, doves, flamingos, ducks, and tortoise livers. The expedition had been requested to collect land tortoises in order to determine their systematics and distribution, so a great deal of time was spent stalking, hauling, and skinning those giant reptiles, Williams doing his share when not collecting or rearing insects. They took 266 tortoises as well as a great many birds and lizards as specimens. This seems a considerable slaughter, but it should be remembered that at this time tortoises and other native animals were being devastated by visiting whalers and by the settlers, and goats and other introduced animals were destroying habitats rapidly. The work of the expedition provided basic knowledge on the fauna that ultimately led to successful conservation efforts.

Williams's collection of insects included more than 150 species of beetles, nine kinds of hawkmoths, five kinds of butterflies, and a diversity of other insects—a much more meager assortment than might have been taken at the same latitude on the mainland, as he noted. Some of the insects Williams worked over himself, publishing on the butterflies and hawkmoths in 1911 and on the bees and wasps in 1926. The latter he found "a very commonplace lot," with a total absence of social species so abundant in Ecuador, only 600 miles away, perhaps because the islands did not supply adequate amounts of insect food to sustain their colonies. The beetles from the expedition were studied by VanDyke, whose posthumous monograph appeared in 1953.

Back at Stanford, Francis continued his studies under Professor Vernon Lyman Kellogg, receiving his degree in 1908. Kellogg had been trained at the University of Kansas and had spent a year working with J. H. Comstock at Cornell. When he moved to Stanford he invited Comstock to teach several winter terms there. Kellogg was a specialist on lice, but he published several widely read general books on zoology, entomology, and evolution. He was an ardent hiker and member of the Sierra Club, where he became a close friend of its founder, John Muir. Later in life, Kellogg became an assistant to Herbert Hoover in the American relief efforts to Belgium, Poland, and Russia. He was also active in organizing the National Research Council, of which he became permanent secretary. All of which proves that the study of lice need not limit one's vision!

But to return to Williams after still another sidetrack. Probably at Kellogg's suggestion, Francis went to the University of Kansas to study for a master's degree. It was here that he began the studies of solitary wasps that absorbed him for much of the remainder of his life. The publication resulting from his research, *A Monograph of the Larridae of Kansas*, nearly one hundred pages long, was devoted to both the systematics and natural history of those insects and was illustrated with the carefully detailed and attractive sketches that became characteristic of most of his later publications.

"Those of us who have not had the good fortune, the patience or the inclination to watch one of these digger wasps at work," Williams wrote, "have missed the opportunity of observing an insect of remarkable instincts, great perseverance, and notable temerity in attacking its often huge prey. Few persons have any idea of the vast amount of good done by these Hymenoptera, for the noxious insects destroyed by the solitary wasps is very great, and plays an important part in maintaining the balance of nature."

From Kansas Williams went to Harvard to study with Professor William Morton Wheeler, receiving his D. Sc. degree in 1915. Here he allowed himself to become diverted from his life's work, studying the light organs of fireflies, which made up a major paper in *Journal of Morphology.* At this time the gypsy moth was a major problem in New England, and Williams joined the staff of the U. S. Department of Agriculture's laboratory at Melrose, Massachusetts, for a brief period. Then, at Wheeler's recommendation, he was offered and accepted a position at the Hawaiian Sugar Planters' Association (HSPA) Experiment Station. With a small loan from Wheeler, he returned to San Francisco to visit his family, then left for Honolulu. He was thirty-four, and remained with the HSPA until he retired at sixty-five.

Williams's task was to assist in the search for natural enemies of pests of sugarcane, a job for which he was uniquely qualified. After only a few days in Hawaii he was sent to the Philippines, where he remained for over a year, studying and shipping back to Hawaii a number of predators of agricultural pests. He must have kept very busy indeed, for he collected more than 180 species of wasps and studied the biology of many of these. His 186-page monograph, *Philippine Wasp Studies* (1919), has rarely been surpassed in the excellence of the descriptions of the behavior of diverse tropical wasps. So accurate and attractive were his drawings that many have been reproduced in other publications. Williams was back in the Philippines again in 1920, where he continued his studies. On both occasions he was located at the Philippine College of Agriculture at Los Banos, where he took advantage of the entomological expertise of its dean, Charles Fuller Baker. Baker was a sufficiently interesting person that he is worth another digression from our major theme.

Baker was trained at Michigan State College, where he graduated in 1892. Already, by graduation, he had amassed an insect collection large enough to occupy several hundred boxes. As assistant to C. P. Gillette at Colorado Agricultural College, he continued to add to his collection from the little-studied Rocky Mountain area. But his devotion to his personal collection was not appreciated by the administration, and he left abruptly, taking everything with him. After that he moved about a good deal, even spending a year studying under Kellogg at Stanford a few years before Williams went there. Service in Colombia and in Cuba enabled him to add a great many tropical insects to his collections.

In 1912 C. F. Baker was offered a position at the Philippine College of Agriculture, where he remained for the rest of his life. He had a laboratory built in his home, with ample space for his large collections, and in the course of time added a great deal to knowledge of Philippine insects. He was also responsible for many improvements in agricultural practices in the islands. He was long estranged from his wife, but became much attached to his cook, Tome San, and intended to marry her as soon as he could get a divorce. But one day his Japanese gardener went on a rampage and tried to kill Baker with an axe. Tome San ran to protect Baker and received the blows of the axe, while Baker escaped unharmed. She died a few days later. Baker never recovered from this episode. He died in 1927, having asked in his will that his ashes be buried in a Buddhist temple next to those of Tome San. His collections were willed to the U. S. National Museum. A representative of the museum, R. A. Cushman, went to the Philippines and took fully five months to pack and ship the 1,400 Schmitt boxes of insects to Washington.

One of the more intriguing by-products of Williams's life in the Philippines was his 1928 paper *The Natural History of a Philippine Nipa House*. He lived for some time in such a house, built of timber, bamboo, rattan, and palm thatch. No doubt modern houses are more comfortable, he remarked, "but they exclude such interesting household pets as the larger lizards, the banana frog, and the several bats, as well as the greater bulk of winged insects that are attracted to lights." One night, when he was writing up his field notes by a lamp, he heard a noise above him and looked up to see a huge cobra stalking rats in the thatch.

The HSPA sent Williams to many parts of the world in the search for predators and parasitoids of use in biological control, including Australia and various parts of South America. He continued to study wasps wherever he went, and the result was another monograph, *Studies in Tropical Wasps* (1928). This is filled with information on species that had been little studied to that time—and many have not been studied since. A few years later Williams compiled a *Handbook of the Insects and Other Invertebrates of Hawaiian Sugar Cane Fields,* a well-illustrated compendium of information on all groups of insects of interest to the sugar cane industry. Somehow he also found time to produce a series of pioneering reports on the aquatic insects of Hawaii. This aspect of his research has been vividly described by Elwood C. Zimmerman, who initiated the important series *Insects of Hawaii,* beginning in 1948:

> *Williams's long years of bachelorhood and his being unburdened by care of property or family responsibilities made possible his spending an unusual amount of time in the pursuit of entomological investigations. He made many trips into the Hawaiian fields and mountains, particularly during his long researches on the aquatic insects. Most weekends found Williams in the field. . . . I accompanied Williams at various times, and learned how careful and thorough were his methods of study. I have considered his discovery of the highly unusual terrestrial larva of the damselfly* Megalagrion oahuense *(Blackburn) to be the crowning achievement of his Hawaiian field studies. Williams had been collecting data for his classic series* Biological Studies in Hawaiian Water-Loving Insects . . . *and he had studied the life histories of all of the Odonata on the island of Oahu with the exception of this fine, large, endemic damselfly. Although the adults could be found without great difficulty, Williams was unable to discover larvae in any of the known or suspected odonate habitats. Over and over he searched for larvae in the areas where the adults flew, but all his attempts were failures. He noted that the adults were often to be seen on steep mountainsides far from water, and week after week he studied their movements. He found that the*

females were sometimes seen on dense thickets of the pestiferous stag-horn fern, Gleichenia linearis, *which forms tangled masses in openings in the lower and middle forests. By persistent observation he found that some females descended into the masses of ferns. Armed with this information, he began a search of the damp trash and leafmold beneath the fern thickets. With extraordinary patience and after repeated and laborious searches that would have defeated a lesser man, Williams finally succeeded in discovering that the nymphs of this damselfly are highly unusual because they are terrestrial and capable of living far from water. The nymphs are very hairy, and their gills are shortened, thickened and hairy and are adapted for a life in the damp ground litter beneath dense masses of vegetation. Most of his other studies on the Hawaiian water-loving insects are the first and only of their kind in Hawaii.*

In 1939, at the age of fifty-seven, Francis married Louisa Clark, of Honolulu, and Louisa accompanied him on further trips, this time to New Caledonia and to East Africa. While in Hawaii, they lived in a cottage surrounded by trees and flowers, designed to attract as many birds and insects as possible. They raised an orphan mynah bird as a pet and together wrote a delightful book about it: *Mike the Mynah*. After Francis's retirement, the couple moved to California, eventually settling near San Diego, where Mary Alice and I visited them in 1954. Even though both were in declining health, they were extremely cordial and happy to "talk bugs" with kindred spirits. Williams was at that time studying the predatory behavior of tarantula hawks in household aquaria, and he was eager to show us his "pets." As to his personal characteristics, I can only confirm the words of E. C. Zimmerman, who knew him much better than I did:

He was a shy, very quiet, extremely modest and retiring person. He had a gaunt appearance that may have resulted from attacks of malaria. He was a lover of classical music, and he played the flute. It was almost unknown of him to speak a harsh word or to reveal any feelings of anger.

Beginning in 1954 and until his death in 1967, Williams and

I corresponded extensively and sent each other specimens relating to our respective research interests. Upon receiving copies of my papers he would invariably write and tell me how much he liked them—an invaluable "shot in the arm" for a struggling young entomologist. Although his eyesight was failing, in his last years he began the study of the systematics of some of the smallest of wasps. His last paper, published when he was eighty-four, dealt with the biology of a spider-hunting wasp he had found nesting in hollow stems in his garden. In characteristic manner, he had found something wholly new, and described it in his usual straightforward style—something only he had eyes to see, even though he was nearly blind at the time!

Williams and Cockerell had very different personalities and had very different life experiences, but they showed a similar excitement about living things. Both were immersed in the world of insects from childhood well into their eighties, and it is doubtful that either had a moment's regret. Both were lucid writers with considerable artistic ability, and their publications, however outdated, read more easily than most published today.

In a letter to me, Francis Williams spoke of the difficulties of "retaining a sufficient part of your personality" in scientific writing. It is true that scientific style has become standardized, dry, and devoid of feeling; anecdotes or unsupported opinions are quickly eliminated by reviewers or editors. This is perhaps inevitable in the present information explosion; it is said that there are more scientists living today than in the entire history of science. But I for one regret the passing of the naturalist as a species of humanity. Cockerell and Williams were two of the best.

18

Confessions of an Incurable Entomophile

Maurice Maeterlinck spoke of the "profound inquietude inspired by these creatures [insects] so incomparably better armed, better equipped than ourselves, these compressions of energy and activity which are our most mysterious enemies, our rivals in these latter hours, and perhaps our successors."

Many decades later, in his book *Lives of a Cell*, Lewis Thomas commented that "the writers of books on insect behavior generally take pains, in their prefaces, to caution that insects are like creatures from another planet, that their behavior is absolutely foreign, totally unhuman, unearthly, almost unbiological. They are more like perfectly tooled but crazy little machines, and we violate science when we try to read human meanings in their arrangements."

The sense that insects belong to a different world than ours is shared by many people, and it is a perfectly valid feeling. After all, the search for a common ancestor of insects and ourselves would take us back more than half a billion years. Insects as a group appeared soon after the land was first occupied, and since that time they have evolved complex relationships with plants and with other kinds of animals. In a sense insects are very much of *this* world, and *Homo sapiens* is a strange and aberrant creature of recent origin who has sought to create his own world, apart from that of nature. However we look at it, insects and humans haven't much in common.

Insects have no capacity to understand humans, so it seems

to me we should make a sincere effort to try to understand *them*. They do, after all, impinge on our lives in many ways, however we may wish it otherwise. In terms of species, they exceed us by at least a million to one. In terms of individuals—one person has estimated that at any one time there are a billion billion insects living on earth. If this is so (and I have no intention of trying to confirm it!), insects exceed humans by a ratio of 200 quadrillion to one. Nothing is gained, and a great deal is lost, by disdaining or avoiding everything that walks on six legs. Entomophobia—fear of insects—is not a condition that permits us to live happily in the real world. In a recent telephone survey of 1,117 households, it was found that only six percent went on record as taking pleasure in insects found outside the home, while over half disliked or were afraid of outdoor insects (88 percent disliked insects *inside* the home).

The revulsion that many people feel is well expressed by Jake Page, writing in *Science 83* (July/August):

> *Most people—I among them—have an altogether understandable aversion to insects. They creep and crawl. They fly at you. They sting, bite, carry diseases, eat odious things, and look awful, especially seen close up. They are the stuff of horror movies.*

I suspect that Mr. Page was not entirely serious (it must be difficult to find material to fill a monthly column!). The point of view he expresses is, however, all too common. But I recommend insects—for their beauty (*especially* close up), for their remarkable behavior, and because we need to understand creatures that *do*, in fact, sometimes bite, sting, and carry diseases. There is something to be said for entomophilia, the love of insects. However, it is perhaps too much to expect everyone to become enamored with creatures so different from ourselves, however remarkable their antics and beautiful their forms and colors. Entomologists may no longer be objects "of pity or contempt," as Kirby and Spence wrote, but their work is cut out for them when they try to bridge the enormous gap between entomophobia and entomophilia. Perhaps it is enough if they sometimes are able to convey a modicum of knowledge of the ways of insects and a respect for the many ways they have found for exploiting the earth's bounty.

It came as a shock to my parents when my early experiences crushing tobacco hornworms between my toes and snagging moths at the lights of my father's fruit market resulted in an incurable case of entomophilia. They had hoped for a business man or, better, a physician, someone able to realize the American dream of making lots of money and surrounding himself with the finest products of western technology. When I eventually took an advanced degree at Cornell University, my father took pains to explain that I was only a doctor of *Philosophy*.

Cornell proved to be a hotbed of entomophiles, many of whom are still good friends. John Henry and Anna Botsford Comstock were gone, but their spirit still lingered, and several of their protégés were still active. The curator of the insect collection was Henry Dietrich, who as a young man had spent a year working for the California Forest Service, where he came to know E. C. VanDyke, whom we met in the preceding chapter. It was here that he acquired his lifelong enthusiasm for beetles. I often "talked bugs" with Hank (as we all called him), but it was not until much later that I learned that he had a family. His wife, Alice, had done research on dragonflies with Comstock's younger colleague, J. G. Needham, and there were also a son and two daughters. Hank had warned his daughters to stay away from entomologists, who were likely to be impecunious and little appreciated by society. Fortunately, his daughter Mary Alice failed to take his advice, and in 1954 married me.

Early in his career, Hank had worked at a variety of jobs, none of which diverted him completely from his passion for collecting insects. For a while he ran a fruit farm on the southern shore of Lake Ontario, where his children were born. Whenever there was a break from farm chores, he was off to the lakeshore or woods to collect beetles. This was not without hazards, for he was a handsome fellow, and he was frequently accosted by a certain local lady who had evidently married for money and not for love. He managed to escape her advances but succumbed to allergies that caused him, after a few years, to accept a job in southern Mississippi as a state plant board inspector.

Southern Mississippi was in those days virgin country for insects, and Hank was often off with his family to local streams, beaches, and woodlands to harvest the fauna. Many of the species they collected proved new to science, and quite a few eventually came to be called *dietrichi* by various specialists (Hank himself never caught the "mihi itch," the hankering to attach his own name to as many species as possible). As a plant inspector, he was often presented with watermelons by farmers, and the Dietrich porch became lined with melons, which they ate from one end of the porch as they added to the other end. Now and then he surprised moonshiners, whom of course he pretended not to notice lest they conduct him away at the point of a gun (one of his favorite collecting sites was called Whisky Creek, with good reason).

As the depression deepened, Governor Bilbo discharged the "yankees," and the Dietrichs bought an old Model A Ford and during the summer of 1932 made their way back toward Cornell. On the way they visited Washington, D.C., and while Hank looked at the national insect collections, Alice and the children toured the museums and government buildings (where the children were often looked at askance by guards for their ragtag appearance and bare feet). When Alice told her three that they would have to wear shoes if they wanted to visit the White House, they demurred. (Even today, I notice that the Dietrich clan shed their shoes whenever they get a chance!)

Back at Cornell, Hank studied for a Ph. D. degree, doing a thesis on the click beetles of New York. During the academic year he was a teaching assistant, and during the summers he roamed the state surveying streams, or on another occasion putting out traps for Japanese beetles. The depression years were not easy ones in which to keep a family of growing children together. For a while Hank participated in a microbiological study of human feces, as a donor, and was able to trade a daily stool sample for a quart of milk. Living as they did in the poorer section of town, his children picked up infestations of head lice at school. Their mother duly treated them with carbolic acid and wrapped their heads in towels. Mary Alice recalls that the minister paid them a surprise visit one day while they were being treated. Head lice were evidently not in his purview. "Playing Bedouin, I see" was his remark.

But times improved, and after he received his degree, Hank

accepted a position involving control of the Dutch elm disease, a beetle-borne scourge that was just beginning to devastate the elms of the northeastern states. Living for several years not far from New York City, he became a member of the New York Entomological Society, where he hobnobbed with a galaxy of insect enthusiasts. Among them were William T. Davis, the Staten Island naturalist; Lucy W. Clausen, whose book *Insect Fact and Folklore* I have sometimes cited; and William Procter, grandson of the founder of Procter and Gamble and for many years director of that company (Mary Alice remembers her thrill at meeting a real millionaire!). Procter had developed an interest in insects as a youth, but it was not until he was forty-five that business duties permitted him to undertake graduate work at Columbia to broaden his training in biology. His summer home on Mount Desert Island, Maine, became a mecca for entomologists, and he organized a biological survey of that island. In the lavish volume on the insects he published in 1946, 6,578 species were recorded from the island, all identified by specialists. Unfortunately, the days of wealthy amateurs are gone.

It was in 1939 that Hank was appointed curator of insects at Cornell. Winters were spent identifying and arranging, summers afield with a net and collecting tubes. He often had to explain what he was doing to passersby (the fate of all entomologists). While working from a state car in the Catskills, he was approached by a lady hiker who, upon being enlightened, exclaimed that "it was nice of the state of New York to hire someone to pick the bugs off the mountains." On another occasion, when collecting along a trail in the Smokies, he was asked where the trail went. "All the way to Maine," he casually answered, "but most people don't go all the way at once." It was, of course, the Appalachian Trail. Hank's remark was typical of a sense of humor that served him well through hard times as well as throughout a long career of doing strange things such as "bug catching." He and Alice eventually roamed much of North America, all the way from the deserts of Arizona to the Canadian Rockies, where they stalked *Grylloblatta*, a "living fossil" halfway between a cricket and a cockroach, that has evidently survived through the ages by hiding beneath glacial ice.

Marrying into an entomological family was the most natural thing in the world for me. Our first home, a modest house in a bramble patch outside of Ithaca, became the site of many studies of wasp behavior, and I wrote a book about it (*Wasp Farm*, 1963). Mary Alice and I have spent every summer in the field studying insects, accompanied by our three children, when they were younger (requiring us to carry five nets!), and often by students. On more than one occasion we have made summers out of winters by going to the tropics or to Australia.

For many years our major goal was to find, study, and photograph *Bembix* sand wasps. It was Henry and Alice Dietrich who first told us of a population of these wasps they had seen at Mammoth Hot Springs, in Yellowstone National Park. It was a species normally found much farther south, but like many other insects it was able to flourish in Yellowstone because of the warming effects of the thermal areas. So we went to Yellowstone, and sure enough, the wasps were nesting along the trail to the springs and even within the circle of benches around Old Faithful. These were scarcely places to conduct research, so we searched for and found places off the beaten track where we could work. We discovered, too, that there was an excellent field station not far away, in Jackson Hole, and we have spent many an idyllic summer there since. The banks of the Snake River, in particular, provided a bonanza of insects deserving of study, and it was a joy to look up now and then to see a bald eagle, or to watch a moose crossing the river. Of course we also took occasion to explore the glorious back country of Yellowstone and the Tetons.

Over the years we have spent time in a number of field stations, which provided not only living quarters and laboratory space, but association with other field-oriented biologists. The Southwestern Research Station, in the Chiricahua Mountains of southeastern Arizona, hosted a particularly interesting group when we were there in the summer of 1959. Theodosius Dobzhansky and his wife were studying *Drosophila* chromosomes; E. Gorton Linsley of the University of California was studying bees and robber files; T. C. Schneirla of the American Museum of Natural History was tracking army ants; and C. P. Alexander of the University of Massachusetts had just surpassed Linnaeus in the number of species he had described.

British biologist H. B. D. Kettlewell, known for his experimental studies of mimicry and protective coloration, also paid a visit. Tarantula hawk wasps abounded in the area, and Kettlewell was especially interested in several large, blue-bodied and orange-winged insects that evidently were mimics of tarantula hawks. These mimics (which included flies, beetles, and ichneumons) were all relatively defenseless insects that apparently gained protection from predators by copying colors and behavior of the same models, wasps that are armed with excruciatingly painful stings. Like a typical Britisher, Kettlewell went about in shorts and sneakers, despite our warnings about rattlesnakes. I remember him standing in a patch of milkweed that was teeming with tarantula hawks. When one of the large, orange-winged insects landed on his bare leg and began walking up toward more delicate parts of his anatomy, he looked down and remarked quite casually: "I say, is that a model or a mimic?" That was carrying entomophilia about as far as it can be carried!

Speaking of tarantula hawks (and why not?—they are spectacular creatures), on a number of occasions I collected these wasps in the Southwest and in Mexico, followed by a group of urchins who asked questions and tried to help. My trick to get rid of them was to pick a tarantula hawk off the flowers with my fingers and show it to them. Of course I always picked up a male, which cannot sting. But my curious followers would pick up a big one, usually female, and quickly decide they wanted no more of that. Once, in Mexico, I especially wanted to collect a series of a particular, poorly known tarantula hawk, so I paid a lad to help me (something I rarely do). Using only his sombrero, he collected an incredible number, suffering innumerable stings for a few pesos. Having been stung so many times myself over the years, I suppose I have developed a lack of sympathy with people who are stung. One's appreciation of nature is never more acute than when a bit of nature is injected into one's flesh.

This last episode occurred at the Pyramids of Teotihuacan, northeast of Mexico City, a place we have visited several times not only for its archeological wonders but for the abundance of insects that swarm about the bushes. Unfortunately, the area became a little more "developed" at each visit, with a corresponding decline in the insect populations. We often wonder if development has struck a particular beach on the west coast of Mexico, where we spent a day netting butterflies beyond belief and enjoying the warm waters of the Pacific—without seeing a soul.

In Cuernavaca, where we lived for several months, great white morpho butterflies crossed our backyard, and our maid Oliva became adept at catching them. When we took her with us to the slopes of Popocatepetl to take care of the children (who were then quite small), she spent most of her time collecting insects and left Mary Alice with the children. We have since wondered what became of Oliva (who was illiterate); at least she tasted the pleasures of entomology for a brief period.

It was *Bembix* wasps that first led us to Australia, for William Morton Wheeler had reported a species preying on damselflies at Lake Violet, deep in the interior of Western Australia. Since all the *Bembix* we had studied in North America preyed on true flies, it seemed worthwhile checking out Wheeler's record—and, of course, every biologist dreams of visiting Australia, a continent so long isolated that it has evolved many unique plants and animals. Along with Bob Matthews, of the University of Georgia, I made my way to Lake Violet, arriving on September 22, 1969, to find that it was no more than a broad, white salt pan that had been dry for some years. We camped there for several days on the brick-red sand, in one of the hottest and most desolate places on earth. Each morning we were awakened by the complex and exceedingly melancholy song of a butcher bird, one that I have tried many times to recall, without success.

The *Bembix* we were looking for was not in evidence, and of course there were no damselflies, which have aquatic larvae. But by digging deeply into the sand we were able to find cocoons surrounded by the wings of damselflies, and by watering the cocoons we were able to cause the wasps to emerge. Evidently, the damselfly larvae are able to survive long dry periods by burrowing beneath the crust of the salt pan; when waters fill the lake they emerge and produce a crop of adult damselflies. The wasps, too, must remain in their cocoons for long periods, em-

---

Tarantula hawks, though formidable creatures, have been objects of special attraction to the author.

P. EADES

erging in response to those same rains.

It turned out that Australian *Bembix* wasps prey on a variety of insects besides flies. One of our most startling revelations came at Ku-ring-gai Chase, a national park just north of Sydney. We were camped with the family in a ravine that reverberated each morning with a chorus of kookaburras; across the stream was a bank that harbored a blue kingfisher with a great orange bill. There were *Bembix* nesting nearby, and they were carrying something small and black into their nests. When we dug up some nests we discovered that they were provisioned with stingless bees, mostly males that the wasps evidently snagged from swarms hovering about the nests of those highly social insects. Later we found other species of *Bembix* using bees as prey. Bee-wolves, of the kind discussed in Chapter 10, do not occur in Australia, and species of *Bembix* have evidently evolved to fill that niche. We also found *Bembix* preying on such things as ant lions, owlflies, lacewings, and even an occasional dragonfly.

On a later trip to Australia I was lucky enough to be able to combine forces for a time with Owain Richards, whom I had come to regard as one of the finest entomologists of our time. Richards had come from London to study the social wasps of Australia, and I showed him a large paper nest I had spotted in a small gum tree. Unfortunately, several branches passed through the nest, so we went out late one night and trimmed off most of them, planning to return the following night and clip the last few so we could capture all the wasps and study the nest. When the time came, Mrs. Richards held the flashlight, I climbed the tree with the clippers, and Owain held the sack. I'm not quite sure what went wrong, but the wasps became thoroughly aroused, and we were stung so many times that our fingers became too swollen to properly dissect the nest once we had collected it. Rather embarrassing—two wasp authorities bungling the job! But eventually we did study the nest, finding 9,140 wasps in it (at least half had escaped). There were seventeen combs containing a total of 130,290 cells, so over time the nest had produced an enormous population of wasps.

(Richards and I usually seemed to become involved in ridiculous situations. Once in Colombia we spent part of a day at the edge of an abandoned orange orchard, throwing oranges at an unusual nest high in a nearby tree, hoping to cause the wasps to attack us so that we could catch some and find out what species was responsible for the nest. We finally succeeded, fortunately without anyone seeing us engaged in so absurd an operation.)

Mary Alice and I visited Australia several times and became much enamored of the continent, its people as well as its flora and fauna. In our 1983 book *Australia: A Natural History*, we interwove some of our personal experiences with an overview of the plants, animals, and human history of the continent, as seen by itinerant Americans. On one of our last excursions, we visited Cairns, in tropical Queensland, where we realized a couple of dreams: to see a bird of paradise (a Queen Victoria rifle bird) and to catch a birdwing butterfly (which we released after reminding ourselves how A. R. Wallace had reacted to the capture of his first birdwing, as recounted in his book *The Malay Archipelago*). A series of ponds nearby hosted a great variety of Australia's superb dragonflies. (I have never quite cured Mary Alice of a trait acquired from her mother: of running off in pursuit of a dragonfly in preference to a wasp.)

I could prolong these anecdotes *ad nauseum* (or have I already passed that point?). But I have overstressed the collecting and field study of insects, as if that were all of entomology. Of course it is not, and in the half-century or so that I have been infected with entomophilia a great many exciting things have happened, a very few of which I have touched on in earlier chapters.

When I was a student, hardly anything was known about insect hormones, and the word pheromone had not even been invented. Just in the last few decades knowledge has mushroomed in fields such as plant-insect coevolution and insect thermoregulation. Much of the skepticism that surrounded the concepts of protective coloration and mimicry has been dissipated by new, experimental approaches to those subjects. The study of behavior has become much more sophisticated, and the science of sociobiology has been born, with the social insects as key elements. Insect control has matured, too, and integrated pest management has emerged to make use of the findings of physiologists, ecologists, and behaviorists to forge scenarios in which the effects of pest species may be mitigated without deluging the earth with poisons. It has been exciting to have known some of the people involved and to have watched the emergence of new revelations, new concepts, new paths to be followed.

For, of course, there is still much to be done. Progress would doubtless be smoother if there were not a number of unfortunate but traditional dichotomies among researchers. The most conspicuous one is that between those who pursue facts and concepts for their own sake, following up whatever leads they have on whatever animal best suits their purposes; and those who feel one should tackle only problems that are "important" from a human perspective. The split between "pure" and "applied" science is pervasive, but is especially so in entomology because of our preoccupation with the many pest species that compete with us for the earth's bounty. How well will we succeed in containing these insects without input from people who study their classification and evolution, their ways of life? And can entomology, or any science, thrive if cloistered apart from society? It is a dichotomy that ought not to exist. Yet the fact is that students of insects for their own sake often do not thrive in the entomology departments of many state universities, and much of the information needed for insect containment (for example, knowledge of hormones and pheromones) emanates from biology rather than entomology departments. Similarly, "pure" biologists often look down upon persons committed to solving mankind's many problems with insects.

Within both of those camps there are still further dichotomies. Some biologists are committed to solving "how" questions: how, for example, does day length induce migration in milkweed bugs? Others are attracted to "why" questions: why, for example, is the asparagus beetle orange, blue, and yellow and why does it lay its eggs erect in rows; what does this mean in terms of adaptation to its environment? Applied entomologists, too, fall into two camps. Do entomologists constitute a rescue squad, dedicated to suppressing outbreaks of pests by the quickest and most effective measures (usually chemical); or are they wise (or perhaps starry-eyed) counselors, planning long-range programs that may eventually enable us to turn insects against themselves with minimum environmental damage?

But the most serious dichotomy is that between the professional entomologist and the layman who thinks of insects (if he thinks of them at all) as something unworthy of more than a swat or a squirt. Entomology was born among laymen, among amateurs such as Kirby, Spence, Darwin, Wallace, Fabre, Melsheimer. It passed to the hands of people only marginally employed in entomology, figures such as Say, Peck, Harris, and Cockerell. Just a few decades ago, it was not unusual for amateurs to attend meetings of many of our entomological societies, but few still do. To a degree all science has become so sophisticated that untrained persons can contribute little. But this is less true of entomology, when there is so much to be learned about even common species. And what of the pleasures of entomology—are those to be left only to Ph. D.s?

I hope not. As an antidote to our almost unbearable social dilemmas, I recommend days afield with an insect net, collecting jars, a camera, and a notebook; and a home equipped with a microscope, rearing cages, and a library of good books on insects. Edward H. Smith, an applied entomologist and former head of the entomology department at Cornell, recently had this to say (in an article in the *New York Times*, August 7, 1983):

> *To a confirmed insect watcher, it amazes me that the sport doesn't have more devotees. We are envious that 700 species of birds in this country have drawn millions of bird watchers, while 90,000 species of insects have a limited following. The potential is there.*

The greatest of all sources of pleasure is *discovery*. Given a plot of earth, whether in a suburban garden, a prairie, or a rain forest, it will be found to be crowded with insects. There are, after all, many levels of discovery. The first seasonal report: the first swallowtail of spring, crossing the backyard. The first personal discovery: perhaps a two-spotted ladybeetle devouring rose aphids. The first regional record: perhaps a Carolina mantid in Montana. Or, of course, something wholly new to science: a new host relationship, new insights on behavior, fresh knowledge on details of life histories. Most estimates have it that only about half the existing species of insects have yet been named and described; most of these are in the tropics, but not all. As we have seen, there are currently intensive research programs on such commonplace insects as hornworms, blister beetles, and honey bees. The air and the bushes are full of wholly unstudied insect species. There seems no end of what may still be learned, and all of what we learn will have a bearing on our ultimate success in coexisting with insects.

For persons of a philosophical bent, there are more profound and perhaps unanswerable questions that can be asked about insects. Are they conscious of their actions? Do they feel pain? Do they enjoy sex? Will they survive a nuclear war and inherit an earth that we failed to fully appreciate? Jonathan Schell entitled the first section of his remarkable book *The Fate of the Earth*: "A Republic of Insects and Grass." Insects (particularly herbivorous species), along with certain kinds of plants, have a much higher tolerance of radiation than most other organisms, including vertebrates. Thus a nuclear holocaust could result in an earth populated by certain plants and insects that might eventually multiply and diversify to fill all the vacant niches. Alas, there would be no one to study them!

Long before the nuclear age, in 1903, W. J. Holland, an "old-time naturalist" associated with the Carnegie Museum in Pittsburgh, speculated on what might happen if life on earth eventually perished because of the death of the sun. This is how he concluded his *Moth Book*, and it will do very well to conclude this book, too:

> When the moon shall have faded out from the sky, and the sun shall shine at noonday a dull cherry-red, and the seas shall be frozen over, and the ice-cap shall have crept downward to the equator from either pole, and no keels shall cut the waters, nor wheels turn in mills, when all cities shall have long been dead and crumbled into dust, and all life shall be on the very last verge of extinction on this globe; then, on a bit of lichen, growing on the bald rocks beside the eternal snows of Panama, shall be seated a tiny insect, preening its antennae in the glow of the worn-out sun, representing the sole survival of animal life on this our earth—a melancholy "bug."

References

Chapter 1: The Pleasures of Entomology

Allen, Garland E. 1978. *Thomas Hunt Morgan. The Man and His Science.* Princeton, N.J.: Princeton University Press. 447 pp.

Dow, Richard P. 1913. The rector of Barham and his times. *Bulletin of the Brooklyn Entomological Society,* vol. 8, pp. 68–74.

Dethier, Vincent G. 1962. *To Know A Fly.* San Francisco: Holden-day. 119 pp.

Frisch, Karl von. 1967. *A Biologist Remembers.* Translated by Lisbeth Gombrich. Elmsford, N.Y.: Pergamon Press. 200 pp.

Teale, Edwin Way. 1949. *The Insect World of J. Henri Fabre.* New York: Dodd, Mead & Co. 333 pp.

Chapter 2: The Lovebug

Hieber, C. S., and J. A. Cohen. 1983. Sexual selection in the lovebug, *Plecia nearctica:* the role of male choice. *Evolution,* vol. 37, pp. 987–92.

Kirkpatrick, Dick. 1973. The truth about those hated lovebugs. *Audubon,* vol. 75, no. 5, pp. 136–38.

Thornhill, Randy. 1976. Reproductive behavior of the lovebug, *Plecia nearctica. Annals of the Entomological Society of America,* vol. 69, no. 5, pp. 843–47.

Chapter 3: The Flea

Cowles, Virginia. 1973. *The Rothschilds. A Family of Fortune.* New York: Alfred A. Knopf. 304 pp.

Rothschild, Miriam. 1965. Fleas. *Scientific American,* vol. 213, no. 6, pp. 44–53.

———. 1983. *Dear Lord Rothschild.* Glenside, Pa.: Balaban Publishers.

Rothschild, Miriam, Y. Schlein, K. Parker, and S. Sternberg. 1973. The flying leap of the flea. *Scientific American*, vol. 229, no. 5, pp. 92–100.

Chapter 4: The Boll Weevil

Adkisson, P. L., G. A. Niles, J. K. Walker, L. S. Bird, and H. B. Scott. 1982. Controlling cotton's insect pests: a new system. *Science*, vol. 216, pp. 19–22.

Cross, W. H. 1973. Biology, control, and eradication of the boll weevil. *Annual Review of Entomology*, vol. 18, pp. 17–46.

Howard, L. O. 1931. *The Insect Menace*. New York: The Century Co. 347 pp.

National Academy of Sciences. 1975. *Pest Control: An Assessment of Present and Alternative Technologies. Volume III, Cotton Pest Control*. National Academy of Sciences, Washington, D.C. 139 pp.

Perkins, John H. 1980. Boll weevil eradication. *Science*, vol. 207, pp. 1044–50.

Chapter 5: The Mormon Cricket

Gwynne, Darryl T. 1981. Sexual difference theory: Mormon crickets show role reversal in mate choice. *Science*, vol. 213, pp. 779–80.

———. 1983. Coy conquistadors of the sagebrush. *Natural History*, vol. 92, no. 10, pp. 70–75.

Wakeland, Claude. 1959. Mormon crickets in North America. *U.S. Department of Agriculture, Technical Bulletin no. 1202*. 77 pp.

Chapter 6: The Gypsy Moth

Doane, C. C., and M. L. McManus, eds. 1981. *The Gypsy Moth: Research toward Integrated Pest Management*. U.S. Department of Agriculture Forest Service Science and Education Agency, Technical Bulletin 1584. 757 pp.

Forbush, E. H., and C. H. Fernald. 1896. *The Gypsy Moth*. Boston: Massachusetts State Board of Agriculture. 485 pp.

Gerardi, M. H., and J. K. Grimm. 1979. *The History, Biology, Damage, and Control of the Gypsy Moth, Porthetria dispar (L.)*. Cranberry, N.J.: Associated University Presses. 233 pp.

Leonard, David E. 1974. Recent developments in the ecology and control of the gypsy moth. *Annual Review of Entomology*, vol. 19, pp. 197–229.

Swain, Roger. 1981. Learning to live with the gypsy moth. *New York Times Magazine*. June 14, 1981.

Chapter 7: "Killer" Bees

Gore, Rick. 1976. Those fiery Brazilian bees. *National Geographic Magazine*, vol. 149, no. 4, pp. 491–501.

Michener, C. D. 1973. The Brazilian honeybee. *Bioscience*, vol. 23, pp. 523–27.

———. 1975. The Brazilian bee problem. *Annual Review of Entomology*, vol. 20, pp. 399–416.

Taylor, O. R., Jr. 1977. The past and possible future spread of Africanized honeybees in the Americas. *Bee World*, vol. 58, pp. 19–30.

Chapter 8: Blister Beetles

Eisner, Thomas. 1974. Cantharidin: potent feeding deterrent to insects. *Science*, vol. 184, pp. 755–57.

Fabre, J. H. 1919. The oil-beetle's journey. In *The Glow-worm and Other Beetles*, translated by A. T. de Mattos. New York: Dodd, Mead, & Co. 488 pp.

Riley, C. V. 1878. On the transformations and habits of blister beetles. *American Naturalist*, vol. 12, pp. 213–19, 282–90.

Selander, R. B. 1964. Sexual behavior in blister beetles (Coleoptera: Meloidae). I. The genus Pyrota. *Canadian Entomologist*, vol. 96, pp. 1037–82.

Chapter 9: The Medfly

Burk, T., and C. O. Calkins. 1983. Medfly mating behavior and control strategies. *Florida Entomologist*, vol. 66, pp. 3–18.

Christenson, L. D., and R. H. Foote. 1960. Biology of fruit flies. *Annual Review of Entomology*, vol. 5, pp. 171–92.

Jordan, W. H., Jr. 1981. Invasion of the medfly. *Natural History*, vol. 91, no. 5, pp. 70–78.

Marshall, E. 1981. Man versus medfly: some tactical blunders. *Science*, vol. 213, pp. 417–18.

Chapter 10: The Bee-wolf Revisited

Gwynne, D. T. 1978. Male territoriality in the bumblebee wolf, *Philanthus bicinctus* (Mickel) (Hymenoptera, Sphecidae): observations on the behavior of individual males. *Zeitschrift für Tierpsychologie*, vol. 47, pp. 89–103.

O'Neill, K. M. 1984. Territoriality, body size, and spacing in males of the beewolf *Philanthus basilaris* (Hymenoptera, Sphecidae). *Behaviour*, vol. 86, pp. 296–321.

O'Neill, K. M., and H. E. Evans. 1982. Patterns of prey use in four sympatric species of *Philanthus* (Hymenoptera: Sphecidae) with a review of prey selection in the genus. *Journal of Natural History*, vol. 16, pp. 791–801.

———. 1983. Body size and alternative mating tactics in the beewolf *Philanthus zebratus* (Hymenoptera, Sphecidae). *Biological Journal of the Linnaean Society*, vol. 20, pp. 175–84.

Simonthomas, R. T., and A. M. J. Simonthomas. 1980. *Philanthus triangulum* and its recent eruption as a predator of honeybees in an Egyptian oasis. *Bee World*, vol. 61, pp. 97–107.

Chapter 11: Marsh Flies

Berg, C. O. 1971. The fly that eats the snail that spreads disease. *Smithsonian*, vol. 2, no. 6, pp. 9–17.

Berg, C. O., and L. Knutson. 1978. Biology and systematics of the Sciomyzidae. *Annual Review of Entomology*, vol. 23, pp. 239–58.

Knutson, Lloyd. 1976. Sciomyzid flies. Another approach to biological control of snail-borne diseases. *Insect World Digest*, vol. 3, pp. 13–18.

Chapter 12: The Milkweed Bug

Dingle, Hugh. 1972. The migration strategies of insects. *Science*, vol. 175, pp. 1327–35.

Dingle, Hugh, ed. 1978. *Evolution of Insect Migration and Diapause*. New York: Springer-Verlag. 284 pp.

Feir, Dorothy. 1974. *Oncopeltus fasciatus*, a research animal. *Annual Review of Entomology*, vol. 19, pp. 81–96.

Rankin, M. A., and L. M. Riddiford. 1978. Significance of haemolymph juvenile hormone titer changes in timing of migration and reproduction in adult *Oncopeltus fasciatus*. *Journal of Insect Physiology*, vol. 24, pp. 31–38.

Chapter 13: The Tobacco Hornworm

Beckage, N. E., and L. M. Riddiford. 1978. Developmental interactions between the tobacco hornworm *Manduca sexta* and its braconid parasite *Apanteles congregatus*. *Entomologia Experimentalis & Applicata*, vol. 23, pp. 139–51.

Heinrich, Bernd. 1984. *In a Patch of Fireweed*. Cambridge, Mass.: Harvard University Press. 194 pp.

Heinrich, Bernd, ed. 1981. *Insect Thermoregulation*. New York: John Wiley & Sons. 328 pp.

Safranek, L., and L. M. Riddiford. 1975. The biology of the black larval mutant of the tobacco hornworm, *Manduca sexta. Journal of Insect Physiology*, vol. 21, pp. 1931–38.

Truman, J. W., and L. M. Riddiford. 1973. Hormonal mechanisms underlying insect behaviour. *Advances in Insect Physiology*, vol. 10, pp. 297–352.

Chapter 14: Enjoying Insects in the Home Garden

Ahmad, Sami, ed. 1983. *Herbivorous Insects*. New York: Academic Press. 257 pp.

Carr, Anna. 1979. *Rodale's Color Handbook of Garden Insects*. Emmaus, Pa.: Rodale Press. 241 pp.

Ehrlich, P. R., and P. H. Raven. 1967. Butterflies and plants. *Scientific American*, vol. 216, no. 6, pp. 104–113.

Futuyma, D. J. 1983. Evolutionary interactions among herbivorous insects and plants. In *Coevolution*, ed. D. J. Futuyma and M. Slatkin. Sunderland, Mass.: Sinauer Assoc., pp. 207–31.

Chapter 15: Some Pioneer American Entomologists

Dow, R. P. 1914. John Abbot, of Georgia. *Journal of the New York Entomological Society*, vol. 22, pp. 65–72.

Kastner, Joseph. 1977. *A Species of Eternity*. New York: Alfred A. Knopf. 350 pp.

Ord, George. 1859. A memoir of Thomas Say. In LeConte, J. L., ed., *The Complete Writings of Thomas Say on the Entomology of North America*. New York: Bailliere Brothers. 2 vols.

Weiss, H. B., and G. M. Ziegler. 1978. *Thomas Say, Early American Naturalist*. New York: Arno Press. 260 pp.

Chapter 16: More Early American Entomologists

Dow, R. P. 1913. The work and times of Dr. Harris. *Bulletin of the Brooklyn Entomological Society*, vol. 8, pp. 106–18.

Flint, C. L., ed. 1872. *A Treatise on Some of the Insects Injurious to Vegetation*, by Thaddeus William Harris, M.D. A New Edition. New York: Orange Judd & Co. 640 pp.

Mallis, Arnold. 1971. *American Entomologists*. New Brunswick, N.J.: Rutgers University Press. 549 pp.

Scudder, S. H., ed. 1869. *Entomological Correspondence of Thaddeus William Harris, M.D.* [with a Memoir by Thomas Wentworth Higginson]. Boston: Boston Society of Natural History. 374 pp.

Wade, J. S. 1926. The friendship of two old-time naturalists [Harris and Thoreau]. *Scientific Monthly*, vol. 23, pp. 152–60.

Chapter 17: A Pair of Latter-Day Entomologists

Cockerell, T. D. A. 1935–1948. Recollections of a naturalist. In 15 parts. *Bios*, vol. 6, pp. 372–85; vol. 7, pp. 149–55, 205–11; vol. 8, pp. 12–18, 51–56, 122–27, 193–200; vol. 9, pp. 21–25, 66–70, 117–24; vol. 10, pp. 35–41, 99–106; vol. 11, pp. 33–38, 73–79; vol. 19, pp. 87–96.

Ewan, Joseph. 1950. *Rocky Mountain Naturalists*. Denver, Colo.: University of Denver Press. 358 pp.

Slevin, J . R. 1931. Log of the Schooner "Academy" on a voyage of scientific research to the Galapagos Islands, 1905–1906. *Occasional Papers of the California Academy of Sciences*, no. 17. 162 pp.

Weber, W. A. 1965. *Theodore Dru Alison Cockerell, 1866–1948*. Boulder, Colo.: University of Colorado Series in Bibliography, no. 1. 124 pp.

Weber, W. A., ed. 1976. *Theodore D. A. Cockerell. Letters from West Cliff, Colorado*. Boulder, Colo.: Colorado Associated University Press. 222 pp.

Zimmerman, E. C. 1969. Francis Xavier Williams, 1882–1967. *Pan-Pacific Entomologist*, vol. 45, pp. 135–46.

Chapter 18: Confessions of an Incurable Entomophile

Evans, H. E. 1963. *Wasp Farm*. New York: Natural History Press. 178 pp. (Also published in paperback by Doubleday Anchor Books, 1973.)

Evans, H. E., and M. A. Evans. 1983. *Australia. A Natural History*. Washington, D.C.: Smithsonian Institution Press. 208 pp.

Evans, H. E., and R. W. Matthews. 1975. The sand wasps of Australia. *Scientific American*, vol. 233, no. 1, pp. 108–115.

Richards, O. W. 1978. The Australian social wasps (Hymenoptera: Vespidae). *Australian Journal of Zoology*, Suppl. ser. no. 61. 132 pp.

Index